Science *for* English Language Learners

Science *for* English Language Learners

K–12 Classroom Strategies

Ann K. Fathman and David T. Crowther, Editors

NSTApress

NATIONAL SCIENCE TEACHERS ASSOCIATION

Arlington, Virginia

NATIONAL SCIENCE TEACHERS ASSOCIATION

Claire Reinburg, Director
Judy Cusick, Senior Editor
Andrew Cocke, Associate Editor
Betty Smith, Associate Editor
Robin Allan, Book Acquisitions Coordinator

PRINTING AND PRODUCTION
Catherine Lorrain, Director
Nguyet Tran, Assistant Production Manager
Jack Parker, Electronic Prepress Technician
Linda Olliver, Cover and Book Design

NATIONAL SCIENCE TEACHERS ASSOCIATION
Gerald F. Wheeler, Executive Director
David Beacom, Publisher

Library of Congress Cataloging-in-Publication Data

Science for English language learners : K-12 classroom strategies / Ann Fathman and David Crowther, editors.
    p. cm.
  Includes bibliographical references.
  ISBN-13: 978-0-87355-253-0
  ISBN-10: 0-87355-253-9
  1. English language--Study and teaching--Foreign speakers. 2. Science--Study and teaching--English-speaking countries. I. Fathman, Ann K. II. Crowther, David (David T.)
  PE1128.A2S323 2005
  428'.0071'2--dc22
                            2005023038

NSTA is committed to publishing material that promotes the best in inquiry-based science education. However, conditions of actual use may vary, and the safety procedures and practices described in this book are intended to serve only as a guide. Additional precautionary measures may be required. NSTA and the authors do not warrant or represent that the procedures and practices in this book meet any safety code or standard of federal, state, or local regulations. NSTA and the authors disclaim any liability for personal injury or damage to property arising out of or relating to the use of this book, including any of the recommendations, instructions, or materials contained therein.

PERMISSIONS

To all the students and teachers
who have shown that the worlds
of science and language complement
and enhance each other

Contents

Section I

Parallels in Language and Science Teaching

Chapter 1

Teaching English Through Science and Science Through English 3

Ann K. Fathman, David T. Crowther

Chapter 2

Learners, Programs, and Teaching Practices 9

David T. Crowther, Joaquin S. Vilá, Ann K. Fathman

Section II
Strategies for Planning, Teaching, Assessing, and Extending Learning

Chapter 6

Science Beyond Classroom Walls:
Fairs, Family Nights, Museums, and the Internet......79

John R. Cannon, Judith Sweeney Lederman, Monica Colucci, Miosotys Smith

Section III
Lessons for Science and Language Learning

Chapter 7

Designing Lessons: Inquiry Approach to Science
Using the SIOP Model......95

Jana Echevarria, Alan Colburn

Chapter 8

Lessons That Work: Science Lessons for English Learners.....109

Ann K. Fathman, Olga Amaral

Section IV
Contexts for Classroom Implementation

Acknowledgments

Thanks to Carolyn Kessler and Mary Ellen Quinn for their inspiration and guidance over the years in showing me the benefits of integrating language and content area teaching and to my family for their continual encouragement and support.

Ann Fathman

A special thanks to my very patient and understanding family, Tammi, Tom, Cassi, and Chris, and to the students and staff at Veterans Memorial Elementary School and my colleagues Joaquin Vilá and John Cannon for opening my eyes and accompanying me on this wonderful journey.

David Crowther

Introduction

cience for English Language Learners is a resource for all teachers who work with linguistically and culturally diverse students. A collaborative effort between science and language educators, it provides a wealth of information on teaching science to English language learners (ELLs). We, as editors of the book, come from the very different, but complementary, fields of English language teaching and science education. Sharing ideas has given us the opportunity to better understand the academic needs of students, to develop new teaching strategies, and to integrate best practices for teaching from both fields. These are insights we hope to pass on to our readers.

Purpose and Audience

Science for English Language Learners is for teachers, prospective teachers, and teacher educators. Its purpose is to provide educators with a guide for teaching science to ELLs. We hope that, by using this book, educators will develop expertise in teaching science content and processes, in language development and literacy, and in inquiry-based teaching while getting practical ideas for teaching. We provide information from both fields by

- describing instructional practices in science and language,
- describing effective teaching strategies,
- providing models for lesson and curriculum development, and
- giving an overview of standards development and implementation.

Organization

The book is divided into four sections.

- In Section I, Parallels in Language and Science Teaching, chapters provide an overview of major themes, principles, and practices.
- In Section II, Strategies for Planning, Teaching, Assessing, and Extending Learning, chapters focus on practical suggestions for the classroom.
- In Section III, Lessons for Science and Language Learning, chapters contain design ideas from language and science educators and exemplar lessons from teachers.
- In Section IV, Contexts for Classroom Implementation, chapters contain an overview of science and English proficiency standards, of research and instructional practices, and ways to integrate science, language, and literacy.

The reader can begin at any part of the book. Readers looking for practical ideas for teaching and designing lessons may focus on sections II and III. Readers needing background in the fields of science and ESL (English as a second language) should read sections I and IV. The book as a whole provides information on theory and practice that should be useful to all educators.

Overview of the Chapters

The book is written by teachers, administrators, and teacher trainers of science and English. Each chapter is coauthored by science and language educators who have done extensive work in their fields and who realize the importance of interdisciplinary teaching. By pairing English and science educators as coauthors on chapters, we capitalize on the strengths from both fields and demonstrate the similarities in teaching methodologies that can be used to reach all students.

Section I: Parallels in Language and Science Teaching

Chapter 1

"Teaching English Through Science and Science Through English." Ann Fathman and David Crowther give an overview of central themes that can guide and improve the teaching of science to English language learners.

Chapter 2

"Learners, Programs and Teaching Practices." David Crowther, Joaquin Vilá, and Ann Fathman provide information on English language learners in our schools and the programs provided for them. They also give an overview of science and language learning principles and how these translate into best practices.

Section II: Strategies for Planning, Teaching, Assessing, and Extending Learning

Chapter 3

"Planning Science and English Instruction." Ann Baumgarten and Marie Bacher describe how to incorporate science, language arts, and ESL standards into the classroom. They offer practical suggestions on how to plan, organize, and implement activities based upon standards, teaching and learning strategies, and student background.

Chapter 4

"Strategies for Teaching Science to English Learners." Deborah Maatta, Fred Dobb, and Karen Ostlund discuss strategies teachers can use to help English language learners learn science while improving their speaking, listening, reading, and writing skills in English. They present ideas on how to connect with students, use collaborative learning, and develop language skills and process skills of inquiry.

Chapter 5

"Strategies for Assessing Science and Language Learning." Anne Katz and Joanne Olson give an overview of principles for assessing language learners in science. They describe how to plan assessment, to use it in the classroom, and to provide feedback and improve learning.

Chapter 6

"Science Beyond Classroom Walls." John Cannon, Judith Sweeney Lederman, Monica Colucci, and Miosotys Smith provide ideas on expanding learning beyond the classroom. They describe informal science learning experiences in museums, learning centers, and science centers. They discuss schoolwide experiences such as science fairs, festivals, and family science nights and then provide internet resources for science and language teachers.

Section III: Lessons for Science and Language Learning

Chapter 7

"Designing Lessons: Inquiry Approach to Science." Using the Sheltered Instructional Operation Protocol (SIOP) Model, Jana Eschevarria and Alan Colburn discuss science inquiry, the SIOP Model, and how to blend the two for good science instruction. They finish with a conversation between a science educator and language expert who give their different perspectives on specific science lessons.

Chapter 8

"Lessons That Work: Science Lessons for English Learners." Ann Fathman and Olga Amaral present formats for science lesson plans that incorporate inquiry and language and science objectives. Teachers from elementary, middle, and secondary levels describe successful lessons, and the benefits of these lessons for English language learners are discussed.

Section IV: Contexts for Classroom Implementation

Chapter 9

"Standards for Science and English Language Proficiency." Margo Gottlieb and Norman Lederman describe the development of the National Science Education Standards and English language proficiency standards. They then discuss new language proficiency standards that integrate science and other content area standards with language standards and give implications for teaching.

Chapter 10

"Perspectives on Teaching and Integrating English as a Second Language and Science." Deborah Short and Marlene Thier briefly review the evolution of ESL instruction and science education. They discuss current promising practices that integrate ESL, literacy, and science. Finally, they highlight innovative programs in U.S. schools that offer interventions that improve the science achievement of English language learners.

Appendixes

The chapters are followed by appendixes that include web references for resources, a glossary of science and language terms, and an overview of safety issues for the science classroom.

About the Editors

Ann K. Fathman

Ann K. Fathman is professor of English at Notre Dame de Namur University where she directs programs in English-as-a-second language teaching and English for international students. She received her PhD from Stanford University and BA in foreign language and science from University of California, Davis. Her professional experience includes elementary, secondary, and college teaching of ESL and science, as well as ESL and bilingual program administration and evaluation. She has taught in Europe and Asia and has been a Fulbright scholar in Slovakia. Her research in applied linguistics has focused on factors affecting second language acquisition, assessment, and heritage language preservation. She has had an interest in science and language teaching for many years, and her publications include coauthoring *Science for Language Learners*, published by Prentice Hall, *Elementary Science ESL Workbooks*, published by DC Heath, and *Teaching Science to English Learners*, published by the National Clearinghouse for Bilingual Education.

David T. Crowther

David T. Crowther is an associate professor of science education at the University of Nevada, Reno. He is an editor of *CESI Science*, which is the journal for the Council for Elementary Science International, and associate editor of the *Electronic Journal of Science Education*, which is the longest-running and first online journal of its kind. He is on the advisory board for the National Science Teachers Association's (NSTA) *Science and Children* and was chair of the NSTA Children's Book Council Committee. He has experience teaching at the elementary/middle level as well as biology at the high school and university levels. He has 13 years of teaching experience at the university level, nine of which have been at the University of Nevada, Reno. Previously, he taught at and received his PhD in Science Education from the University of Nebraska—Lincoln. He has published 24 articles that are both research based and practical for elementary science education and has done science education workshops and presentations in 39 states. He is the past president of CESI and a former board/council member of NSTA.

Author Biographies

Olga Amaral

Olga Amaral chairs the Division of Teacher Education at San Diego State University, Imperial Valley Campus. She is also an associate professor in the Department of Policy Studies in Language and Cross-Cultural Development. She received her EdD from the University of Massachusetts at Amherst. She serves as the director of the California Science Project in Imperial Valley and is the principal investigator for several grants that promote greater understanding and preparation for teachers of English learners. Her research and publications emphasize the instruction of English learners in the content area. Specifically, she focuses on methodology used in classrooms with English learners. Her publications have focused on improving student achievement for English learners by linking aspects of science instruction and English language development. Through her collaboration with the Valle Imperial Project in Science (VIPS) (see Chapter 10), she has helped to develop training modules for teachers that involve such techniques as lesson study and an integration of both science and English language development (ELD) standards. She has widely disseminated information about this work both nationally and internationally.

Marie Bacher

Marie Bacher is a science resource teacher and a classroom teacher for the Santa Clara Unified School District, California. In her 15 years as an educator she has been a tutor, preschool teacher, an upper-grade multiage teacher, a science camp director, and director of environmental education. She has a masters in education with an emphasis in administration and supervision from San Jose State University. She has spent the last several years developing and implementing a hands-on science curriculum that integrates best practices in inquiry, language arts, and English language learner (ELL) strategies. She frequently does science staff development for literacy specialists, principals, environmental educators, and her fellow teacher colleagues in science inquiry. Her work focuses on strategies for English language learners in science, science notebooks, performance-based assessments, and science process skills. She started her science-teaching career in a residential outdoor science school and to this day believes the best way for everyone to learn is through hands-on experiences.

Anne Baumgarten

Anne Baumgarten is a science/literacy resource teacher with Santa Clara Unified School District, California. She is responsible for designing and delivering staff training on science instruction as well as reading and writing workshops for elementary school teachers. She works with Partnership for Student Success in Science/Bay Area Schools for Educational Excellence, a nine-district consortium of science teachers that provides training in science content and inquiry methodology. She supports the integration of the language arts and science through classroom mentoring in the Guided Language Acquisition by Design program. She has been in education for more than 15 years, teaching adults as well as young children. In addition she has worked as a science writer for the University of Southern California and as a writer of children's educational television programs for Disney Animation. She has an undergraduate degree in science writing from the University of California at Santa Cruz and is currently completing an administrative credential.

John R. Cannon

John R. Cannon is associate professor of science education at the University of Nevada, Reno. His interest in classroom technologies and their applications began in 1987. He holds a PhD in Science Education from Kansas State University, an MA in classroom teaching from Central Michigan University, and a BA in Elementary Education from the University of Montana. In 1996, he launched the first totally electronic professional journal related to science education and research: the *Electronic Journal of Science Education*. In 2000, he researched and developed *Merrill Education's Links to Science Education Resources Website*. His chapter on distance learning in science education can be found in *Evaluation of Science and Technology Education at the Dawn of a New Millennium*, Kluwer Publishers, 2002. His current research interests include second language acquisition strategies and their close relationship to learning elementary science.

Alan Colburn

Alan Colburn is an associate professor of science education at California State University Long Beach. He holds a PhD in Science Education from the University of Iowa, as well as other degrees from the University of Pennsylvania, University of Illinois, and Carnegie-Mellon University. He has taught high school chemistry, advanced placement chemistry, and physical science. He currently teaches undergraduate students, students and teachers pursuing teaching credentials, and graduate students. His interests include not only inquiry-based instruction, but also the nature of science. Recent research compared science teacher and clergy views on evolution, creationism, science, and religion. He has authored 27 publications and given 46 presentations. This is the ninth time his work has appeared in an NSTA publication.

Monica Colucci

Monica Colucci teaches math and science in Miami Dade School District, Florida, to grades three through five. She has 11 years of teaching experience and has worked with diverse student populations, such as English language learners, students with disabilities, and gifted children. She received a BS in Elementary Education and a master's degree in educational leadership from Florida International University. She is certified in the areas of English for speakers of other languages (ESOL) and gifted education. She has served as a teacher consultant for the University of Miami's Science For All for seven years and helped develop and write the instructional units for this project, trained teachers to use the materials in their classrooms, and made presentations on this topic at professional seminars. She works closely with school administrators and teachers to develop and implement schoolwide strategies to enhance the academic performance of students, especially that of limited English-proficient students and students with disabilities.

Fred Dobb

Fred Dobb, PhD, Stanford University, has been director of the English Learner Initiative of the California Science Project (CSP) and has spent his career in language minority programs as a bilingual teacher, administrator, and staff development specialist. He has been California Department of Education director of Bilingual Education, state supervisor of International Language Programs. Before joining CSP, he was a collaborator on the California English Language Development Test. He teaches courses in linguistic and cultural diversity and second language acquisition at San Francisco State University. He is the recipient of the California Language Teachers Association President's Award. A Fulbright scholar in Brazil, he has taught at postsecondary institutions in Puerto Rico, Nicaragua, Mexico, and Spain, and has trained science teachers from Chile and Argentina at the University of California, Davis.

Jana Echevarria

Jana Echevarria, PhD, is chair of the Department of Educational Psychology, Administration and Counseling at California State University, Long Beach, and a professor of Special Education. Her professional experience includes elementary and secondary teaching in special education, English as a second language (ESL), and bilingual programs. She has lived in Taiwan and Mexico where she taught ESL and second language acquisition courses, as well as in Spain where she conducted research on instructional programs for immigrant students. Her research and publications focus on effective instruction for English language learners, particularly those with learning disabilities. She has written numerous journal articles and book chapters, has written and produced two videotapes, and has coauthored two books: *Sheltered Content Instruction: Teaching Students with Diverse Abilities* and *Making Content Comprehensible for English Language Learners: The SIOP Model*, both published by Allyn and Bacon.

Margo Gottlieb

Margo Gottlieb is director of assessment and evaluation at the Illinois Resource Center, Des Plaines, and lead developer for

World-Class Instructional Design and Assessment (WIDA), a multistate consortium devoted to creating an assessment system for English language learners. In that capacity, she framed English language proficiency standards for 10 states. She holds a PhD in Public Policy Analysis from the University of Illinois at Chicago with a specialization in evaluation research and program design. She has authored an array of books, monographs, and articles and has constructed numerous assessment instruments. She has served as a consultant and adviser to numerous states, government agencies, organizations, universities, and publishers. In addition, she is a member of various national and state task forces and expert panels. Active in Teachers of English to Speakers of Other Languages (TESOL), she currently chairs the committee on revising its preK–12 student standards. Experienced in presenting and consulting nationally and internationally, she recently served as a Fulbright senior specialist in Chile.

Anne Katz

Anne Katz has worked for more than 20 years as a researcher and evaluator with educational projects involving linguistically and culturally diverse students. She received a PhD in Second Language Education from Stanford University. As a lecturer at the School for International Training in Brattleboro, Vermont, she teaches courses in curriculum, assessment, and evaluation. She has also worked as a teacher educator in Brazil, Egypt, and Ukraine. She led the TESOL-sponsored team that developed assessment guidelines for the preK–12 ESL Standards, and her most recent publications focus on standards-based assessment systems. She currently serves on the TESOL committee revising student standards. In her work, she promotes linkages between research and the classroom to support student learning and teacher development.

Judith Sweeney Lederman

Judith Sweeney Lederman is the director of Teacher Education in the Department of Mathematics and Science Education at Illinois Institute of Technology. Her experience with informal education includes her work as curator of education at the Museum of Natural History and Planetarium in Providence, Rhode Island. She regularly presents nationally and internationally on the teaching and learning of science in both formal and informal settings. In addition to numerous book chapters, she has recently published an elementary science teaching methods text and is currently writing a secondary methods text and two books on the nature of science. She has served on the boards of directors of NSTA and CESI and is president of CESI.

Norman G. Lederman

Norman G. Lederman is chair and professor of mathematics and science education at the Illinois Institute of Technology. He received a PhD in Science Education and has MS degrees in both biology and secondary education. Prior to his 20-plus years in science education, he was a high school teacher of biology and chemistry for 10 years. He is internationally known for his research and scholarship on the development of students' and teachers' conceptions of the nature of science and scientific inquiry. He has been author or editor of 10 books, written 15 book chapters, published more than 150 articles in professional journals, and made more than 500 presentations at professional conferences around the world. He is a former president of the National Association for Research in Science Teaching (NARST) and the Association for the Education of Teachers in Science (AETS). He has also served as director of teacher education for NSTA and has served on the boards of directors of NSTA, AETS, NARST, and the School Science and Mathematics Association.

Deborah Maatta

Deborah Maatta is a project coordinator with the District of Columbia Public Schools Office of Bilingual Education. She began teaching English as a foreign language in West Africa where she also worked as a technical trainer for the U.S. Peace Corps. She went on to teach content-based ESL at Lincoln Multicultural Middle School in the District of Columbia. She coordinated the "Hands-On Science Program," a Title III project designed to improve science education for middle school level ESL students in the District of Columbia Public Schools. She currently manages a Title III Teachers and Personnel grant and a Refugee Children School Impact Grant. She received an MA in Education from American University.

Joanne K. Olson

Joanne K. Olson is an assistant professor in the Center for Excellence in Science and Mathematics Education at Iowa State University. She received a PhD in Science Education in 1999 from the University of Southern California. She earned a master's degree in education in 1993 from the Claremont Graduate University and received a bachelor's degree in liberal studies with a concentration in science from California State Polytechnic University, Pomona, in 1991. Her research interests focus on science teacher preparation and cognitive issues in the learning of science, including the role of the nature of science. She was an elementary and middle school science teacher in South Central Los Angeles before moving to Iowa. Currently, she coordinates the elementary science methods courses at Iowa State University, and co-directs the master's of arts in teaching program in Science Education.

Karen Ostlund

Karen Ostlund is professor and director of The Center for Science Education at the University of Texas, Austin. Her many honors include the NSTA Distinguished Teaching Award and Alpha Chi Favorite Professor at Southwest Texas State University, 1996. She is a past president of CESI and has served on the NSTA board of directors. Among her many publications are *Rising to the Challenge of the National Science Education Standards: The Process of Science Inquiry, Primary Grades and Grades 5–8,* in two volumes (with S. Mercier), and *Science Process Skills: Assessing Hands-On Student Performance.* She has authored numerous invited chapters, journal articles, and specialty publications. She has been a major contributor to several science textbook series for use at the elementary and middle levels. She has presented at more than 100 workshops across the country.

Deborah J. Short

Deborah J. Short, PhD, is director of the Language Education and Academic Development division at the Center for Applied Linguistics in Washington, DC, and co-developer of the SIOP Model for sheltered instruction. She was coprincipal investigator for a research study on the effects of sheltered instruction on English language learner achievement and directed the national English as a Second Language Standards and Assessment project for TESOL. She currently directs a study on secondary English language learners funded by the Carnegie Corporation and the Rockefeller Foundation and is a senior researcher on a U.S. Department of Education evaluation study of programs for students in grades K–3. She has extensive experience in school-based research on the integration of language and content instruction and on programs for English language learners. She regularly provides professional development to teachers around the United States and abroad. She develops curricula and instructional materials for students and has authored or coauthored numerous publications, including two ESL series, *High Point* and *Avenues*, from Hampton-Brown. She has taught English as a second or foreign language in New York, California, Virginia, and

the Democratic Republic of Congo.

Miosotys S. Smith

Miosotys S. Smith was born in Cuba where she spent the first 15 years of her life, after which she immigrated with her family to the United States where she continued her education. She completed her undergraduate studies at St. Thomas of Villanova University in psychology, received her teaching certification at Florida International University, and earned a master's degree in early childhood education from Nova Southeastern University. During her 17 years of teaching, she has taught grades prekindergarten through five, and had the opportunity to work closely with ELL students and interact with their families. She is certified in gifted education and has been teaching gifted children for the past eight years. In order to encourage parental involvement, she has developed and implemented numerous workshops for parents in the areas of reading and problem solving. She also has sponsored and led schoolwide programs and competitions such as science fairs and spelling bees, oratorical and book-writing contests, Odyssey of the Mind, and Math Bowl.

Marlene Thier

Marlene Thier is a veteran of the classroom, a science materials developer, a teacher educator, and a leader in the movement to link science and literacy education. She has made presentations on the subject at conferences from California to South Africa and has worked closely with the New York City schools to implement a program based on her concepts. She is codeveloper and teacher education coordinator for the Science Education for Public Understanding Program (SEPUP) at the Lawrence Hall of Science on the Berkeley campus of the University of California. She is also cocreator of more than a dozen other inquiry-based science courses and modules for SEPUP. Marlene is a coordinator of SEPUP's Elementary Science Teacher Leadership program, funded by EXXON/Mobil, which develops workshops and printed materials to help preservice and inservice educators teach science more effectively. She has worked as a coauthor on the program's 10 guidebooks on subjects such as curriculum integration and combining math and science.

Joaquin S. Vilá

Joaquin S. Vilá, PhD, is a native of Puerto Rico where he completed his BA in English with an emphasis in linguistics and secondary English education. Upon graduation, he taught ESL in grades 7–12 in both public and private schools. He received both an MA and a PhD in Linguistics from Michigan State University. He has been involved in higher education for close to 20 years in the development, implementation, and administration of ESL teacher-preparation programs and intensive English language programs. His professional pursuits also involve development of inservice training opportunities for school personnel in the areas of reading, assessment, and content-area instruction for ESL learners. His current research interests include content-literacy development for ESL learners, ESL assessment, and professional development for school personnel. He is currently associate professor and ESOL adviser with the English Department at Salisbury University in Maryland where he is actively engaged in teaching graduate and undergraduate level courses as part of TESOL programs, supervising TESOL interns, collaborating in professional development projects, participating in TESOL/National Council for Accreditation of Teachers professional program reviews, and advising students about the rewards and challenges of a teaching profession.

Section 1

Parallels in Language and Science Teaching

Teaching English Through Science and Science Through English

Ann K. Fathman and David T. Crowther

Eduardo came to the United States a little more than three years ago. He spoke little English. After a brief time at an intake center, Eduardo was sent into a regular sixth-grade classroom. He immediately found friends who spoke his native language and translated for him. After all, he was smart and had attended school in his native country—he just didn't understand English. And he had a very supportive family who encouraged him to learn and be successful.

Fortunately for Eduardo, he was in classrooms in which teachers were trained in sheltered instruction, used cooperative learning strategies and lots of hands-on instruction, accommodated different learning styles, and used assessment strategies that allowed him to demonstrate his knowledge of a subject even with his limited command of English. Over time, he became more comfortable in the welcoming environment provided by the teachers and began to understand the new language he was immersed in.

Eduardo was pulled out of his regular classroom for English instruction during his first two years. By his third year, he had learned conversational English, could read and write basic English, and had begun to understand some of the technical aspects of academic English. He began to feel confident in his learning again. As Eduardo's confidence increased, so did his skill. He needed less and less help from his English instructor.

When Eduardo reflects upon his experience in America, he fondly remembers his science class where he worked with real wires, bulbs, and batteries as he constructed a simple circuit. He still is surprised at how this experience both fascinated him and encouraged him. He was able to construct both science knowledge and English language that described what he was learning. He remembers that when the wires were put together in the right order with the battery and lightbulb, the bulb lit up and that made a "complete circuit."

Window *Into the* Classroom

He had greatly enjoyed the hands-on aspect of the lesson, but was really amazed how he had naturally learned about circuits even when his English was very limited. That lesson not only began a lifetime interest in physics, but also taught him that he could learn and be successful in America, a place he now called home.

What better way to learn English than through the study of science, and what better way to learn about science than experiencing it through language and literacy in and out of the classroom. Sutman, Allen, and Shoemaker (1986) observed in *Learning English through Science* that science and language link us to knowledge of the world and beyond, to understanding people, phenomena, and processes. But this understanding is difficult to obtain in our culture without developing proficiency in the English language.

Eduardo's experience in America described in "Window Into the Classroom" is a success story. Through his experiences in school, he was able to succeed. He developed a love for science as he learned English.

We hope this book will be a resource for all teachers who have a responsibility to teach science and wish to provide quality education to linguistically and culturally diverse students such as Eduardo. We describe instructional practices and programs, standards and goals, teaching strategies, and program and lesson design. A number of reoccurring themes emerge from these chapters that are worth noting. We list some of these major themes and the chapters where they can be found to serve as a guide to the book.

Connection Between Science and Language Development

Every science lesson is a language lesson. Inquiry-based science has been found to have many benefits for students who are developing proficiency in English. By merging language and science, teachers can help students learn both more effectively (see Chapter 4).

Engaging in hands-on experiences in science provides opportunities to engage English language learner (ELL) students, but expressing an understanding of science concepts inherently requires the use of language, and science is language intensive. ELLs face two learning tasks: they need to understand the science content in the lesson and the language associated with that content (see Chapter 5).

Research suggests that science can enhance the language development of children with limited English (see Chapter 10).

Addressing the Needs of Diverse Students

English language learners are diverse because they represent different cultures, but also different languages, educational and family backgrounds, and levels of native and English language proficiency (see Chapter 2 and Chapter 10).

In the classroom that affirms linguistic diversity, teachers encourage ELLs to expand their primary language literacy skills, which affirms the value of the student, deepens student understanding of vocabulary, and strengthens literacy (see Chapter 4).

Teachers should be aware of the diverse cultures and language abilities of students when assessing what they know or have learned. Using multiple assessments, providing clear feedback, and setting achievable, yet challenging goals in science and English help students demonstrate their understanding in a variety of ways and monitor their own progress (see Chapter 5).

The goal of "learning for all students" is often compromised by cultural, societal, and language differences. The teacher's role is to create a classroom environment wherein all students feel accepted, encouraged, and empowered to participate actively in learning (see Chapter 7).

Because students come from diverse backgrounds, it may be necessary for teachers to build the skills needed to perform inquiry-based science activities. Scaffolded inquiry can provide essential support as students construct the skills and knowledge needed to build science literacy. Students can pass through a "continuum of inquiry" (direct, guided, full) to learn skills necessary to engage in inquiry (see Chapter 4 and Chapter 7).

Lessons should provide opportunities for guided support from the teacher and help from peers so that all students can participate in activities, irrespective of their level of English proficiency (see Chapter 8).

Impact of Standards on Teaching

Implementing standards involves a dynamic interaction between content standards, language and literacy standards, and the abilities and needs of students (see Chapter 3).

Standards serve as an excellent starting point for designing units and assessments because standards can be used to organize a unit around big ideas students are to learn, not isolated facts (see Chapter 4).

Effective lessons include objectives based upon language standards that are designed to work harmoniously with the science content standards to introduce students to aspects of English language development while they study science content (see Chapter 8).

Educational reform with a focus on standards has had an impact on all teaching. "Science for all" and "science for inquiry" are at the center of science reform efforts. All students should learn how to do inquiry as well as learn the traditional science content (see Chapter 9).

ESL standards have focused attention on English language learners and show how teachers can help these students be successful in mainstream classes. Language standards are being integrated with content standards so that teachers can blend science and language with an integrated approach. Together, English language proficiency and science standards provide a platform for the vision of how ELLs can successfully access the science curriculum (see Chapter 9).

Similarities in Science and Language Learning Processes

Learning science and a language are cognitive processes that support each other. The science process skills—including observing, predicting, communicating, classifying, and analyzing—are almost the same as language learning skills—seeking information, comparing, ordering, synthesizing, and evaluating. These skills are truly the key to integrating content instruction with language acquisition (see Chapter 3).

As a teacher helps students develop the science process skills of inquiry, language process skills or language learning strategies are simultaneously being developed (see Chapter 4).

Two fundamental characteristics of the learning process, transfer and language dependence, frame our understanding of critical issues in teaching and assessing English learners in the science classroom (see Chapter 5).

In the classroom, science and language are interdependent, in part because each is based on process skills that are mirrored in each other. Both science and English instruction focus on skills such as noting details, predicting, distinguishing fact from opinion, and linking words with precise meanings (see Chapter 10).

Overlap of Best Practices

Core curriculum principles for learning and teaching science are similar to those for language. There is overlap of best practices that recommend the use of meaningful activities encouraging hands-on, active, cooperative participation, with connections to the experiential world (see Chapter 2).

Teaching strategies that help students learn English and science simultaneously should be used. These include strategies related to connections with students, teacher talk, student talk, vocabulary, reading, and writing skill development (see Chapter 4).

Extending learning beyond the classroom and involving family and community are common goals of both science and language teaching. These can be achieved through activities such as off-campus visits, science fairs, family science nights, and the use of technology (see Chapter 6).

Inquiry-based science lessons encourage English language learners to use academic English by interacting with peers, which makes the lessons especially effective for language development. The eight components of the SIOP (sheltered instruction observation protocol) approach reflect what we know about effective science teaching and about high-quality instruction for English learners (see Chapter 7).

Good science and language instruction emphasize the teaching of process skills and learning strategies to help students access, analyze, and retain information. Clear objectives for both science and language stated in terms of what students should be able to do should be a part of all lessons. The goals are similar—getting students to think about and understand new concepts and ideas in meaningful ways (see Chapter 7).

Teachers should learn how to scaffold, not only for language, but also for scientific inquiry. Guided inquiry allows students to become engaged, use information to reason through a scientific issue, master concepts, and design their own projects. By embedding inquiry and sequencing investigations, goals for science and literacy can be attained (see Chapter 10).

Importance of Collaboration

Science and language concepts can be taught simultaneously through practices related to each field. Science and language teachers must plan and work together to serve the needs of English language learners (see Chapter 2).

Input from both science and language teachers in creating lessons can ensure that components are included that encourage science inquiry while at the same time building background, providing practice, emphasizing vocabulary, reviewing, and providing assessment for learners at all proficiency levels (see Chapter 7).

Collaboration among teachers of ELLs is key to ensuring an integrated approach. Joint time for planning is essential for teachers to develop standards-based teaching and assessment activities (see Chapter 9).

A relatively new trend is the coteaching model with which an ESL teacher spends part of the day in the regular classroom coteaching with the grade-level teacher (see Chapter 10).

Need for Professional Development

Teachers need to develop expertise in teaching science content and processes as well as in teaching language and literacy (see Chapter 2). The need for teacher expertise in English language development is immediate and widespread (see Chapter 3). With a focus on high-stakes assessment, teachers must have training in how to create a coherent plan to document students' understanding and skills in language and science (see Chapter 5).

Current practice favors content-based program models, but preservice teacher education has not kept pace for elementary and secondary teachers. Most states do not require teacher candidates to take courses in ESL methods or sheltered instruction techniques. There is a continuing need for professional development of teachers (see

Chapter 10).

Science, when done properly and within the constraints of the discipline, truly supersedes physical and political boundaries. Both science and language have global and personal applications and help students learn about the world around them. In the classroom, the worlds of science and language coincide and can enhance and extend each other as is evident in the pages of this book.

Reference

Sutman, F., V. Allen, and F. Shoemaker. 1986. *Learning English through science.* Washington, DC: NSTA Press.

Further Reading

Bybee, R., ed. 2002. *Learning science and the science of learning.* Arlington, VA: NSTA Press. This volume of contributed chapters is based on the latest research in science education. Chapters include crosscurricular and integrated teaching strategies as well as the latest in assessment and learning in science.

Halley, M., and T. Austin. 2004. *Content-based second language teaching and learning: An interactive approach.* Boston: Allyn and Bacon. A thorough and updated examination of interactive approaches in second language instruction. The authors address practical strategies for implementing content-based language teaching and learning. Relevant activities are provided to ensure student comprehension. Each chapter includes refreshing comments from classroom teachers that add relevance to each theme addressed. Though intended primarily for ESL teachers, science instructors will find it both relevant and approachable.

National Science Teachers Association. 2001. *Science learning for all: Celebrating cultural diversity.*

Arlington, VA: NSTA Press. This is a compendium of articles and best practices from NSTA's high school journal *The Science Teacher*. This collection provides fresh ideas on how to meet the learning needs of all students in the science classroom. The book covers three must-know areas of multicultural science education, curriculum reform, and teaching strategies in science and language and provides practical insights into how to give students an appreciation of the contributions that all cultures make to our scientific knowledge.

Learners, Programs, and Teaching Practices

David T. Crowther, Joaquin S. Vilá, and Ann K. Fathman

Mr. Rocco has taught science for five years at an urban middle school that has recently received an award for excellence in multicultural education. This school has a very diverse student body. In addition to science training, Mr. Rocco has had professional training and certification in teaching English learners and is bilingual in English and Spanish.

He has taught different kinds of classes serving English learners. He has taught small groups of English learners in pullout classes designed to help them with English and content area classes. He has taught science bilingual classes composed of only Spanish students in which he used Spanish to help his students understand key science concepts. He has taught biology in sheltered science classes containing students from different countries where he spoke only English but used special strategies to make science more understandable to his students. And he has taught regular science courses and labs containing students with varying abilities and backgrounds.

He has become aware over the years that there is neither "one best way" to teach nor "one best program." In all the classes that Mr. Rocco has worked with, students have benefited as he has helped them learn English while they were studying about science. He is aware of how the abilities and needs of the students change from year to year and how the strategies he uses to help them have to be continually reassessed. He is continually amazed at how so many of his students have successfully learned English while being engaged in hands-on, inquiry-based science activities.

Window *Into the* Classroom

The need to educate children from diverse backgrounds is growing. The changing demographics of the United States and recent focus on educational standards have made it increasingly critical to address the needs of the many linguistically and culturally diverse students in our schools. Schools throughout the United States are attempting to meet the needs of this diverse population through a wide variety of programs, such as those taught by Mr. Rocco, the teacher of science to English language learners in the "Window Into the Classroom." The overview in Chapter 1 provides a clear indication that teaching practices founded on sound theory and reflective of professional standards are indeed doable tasks. Mr. Rocco's experience is a clear and valid testament to that.

Diversity in Schools

Our classrooms are filled with students from different races, cultures, religions, socioeconomic levels, abilities, and learning styles. All of these differences and many more encompass the diversity in our schools. Recent years have brought an increasing focus on multicultural education to accommodate the diversity of students represented in the classroom. The importance of multiculturalism is recognized in positions taken by all national teacher organizations, including the National Science Teachers Association (NSTA) and Teachers of English to Speakers of Other Languages (TESOL), as shown in the following statements of purpose:

We believe that all children can learn and be successful in science and our nation must cultivate and harvest the minds of ALL children and provide the resources to do so. (NSTA 2000)

The purpose of contemporary education is to prepare all students for life in the world, including those learners who enter schools with a language other than English. (TESOL 1997)

One of the many facets of multiculturalism is illustrated by children who enter schools speaking a language other than English. Knowing how to educate these children is no longer the mission of only special teachers or aides but is now important for all teachers. With the advent of the No Child Left Behind Act, the regular classroom teacher, the subject specialist, and administrators all need to understand how English is learned as a second language and the teaching strategies that accompany and facilitate learning for these students if a school is to meet the standards the act mandates.

Although in the past, language and content area instruction followed distinct paths, the need for integration of these has become increasingly evident for science- and second language-teaching professionals. The fast-paced changes in school populations across the nation are a catalyst for such a shift in instructional vision.

The statistics are indeed astounding—the number of students entering our schools with limited English has increased significantly in recent years and continues to rise. According to the U.S. Department of Education, during 2000–2001 nearly 10% of K–12 public school enrollment was "limited English proficient." This is more than

4,580,000 students. The student population with limited English grew approximately 105% between 1991 and 2001, while the native English speaker population grew only 12%. Approximately 44% of English learners are in grades K–3, 35% in grades 4–8, and 19% in grades 9–12 (Kindler 2002). All these numbers are indicative of the challenges faced by schools around the country.

Figures show that the foreign-speaking population is not distributed evenly among the states. The five states with the largest numbers of English learners are California, Florida, Illinois, New York, and Texas. These states have had an increase of more than 200% in students speaking another language at home over the last 10 years. Not only is the population not distributed evenly among states, but the number of languages spoken in different regions also varies significantly. Spanish, however, is spoken more than any other language in all regions. The U.S. demographic data also suggest that 4.4 million households in 2000 were "linguistically isolated"; they reported no person over age 14 speaking English well (Shin 2003). This situation creates particular challenges for providing meaningful educational opportunities to English language learners, which are compounded because efforts to engage parents as participants in their children's education are also hindered by linguistic and cultural barriers.

The figures cited above are indicative of the progressively rapid rate at which a diverse population is entering our classrooms. Census data, as well as other relevant sources, reveal the need to identify suitable instructional alternatives that would allow schools to face these demographic realities of the present century (Freeman and Freeman 2000).

English Language Learners

English language learners are not a homogeneous group, and a variety of terms have been used to describe them, sometimes causing confusion. One of the earliest and most widely used terms in the United States is *English as a second language* (ESL) learners; another is *English for speakers of other languages* (ESOL). English learners in non-English-speaking countries are frequently referred to as EFL (English as a foreign language) learners. The term *limited English proficient* (LEP) has been used in schools and federal legislation to designate students who are beginners and intermediate in English acquisition, but is being replaced by the less-negative term *English learner* (EL) or *English language learner* (ELL). The latter term—ELL—is the one that will be used most frequently in this book.

As a group, students from linguistically and culturally diverse groups tend to have higher school dropout rates and score lower on achievement tests, including science (Siegel 2002). But it is a misconception that all children coming from immigrant families struggle academically. In fact, many excel. Anderson (2004) says that 60% of the top science students (38% of the U.S. physics team and 25% of the Intel Science Talent Search) are children of immigrants. This shows that many of these students are capable of excelling if provided with good educational opportunities.

English language learners entering schools come from diverse educational backgrounds and have different levels of proficiency in English as well as in their native languages. For example, some students may enter school with little English but be literate in their native language and at grade equivalent in science or math. Others may enter with little formal schooling and not be able to read or write in their native language or in English. Some students appear not to speak a word of English but are able to read and write; others may be fluent in speaking but lack literacy skills; while still others may be at grade level in speaking, listening, reading, and writing English. It is important to know that, even though an ELL student may have conversational English skills, that student may not yet have developed the skills to read, comprehend, or converse in academic English. Assessment of language skills and stage of language development becomes a very important part of educating students with limited English because the needs of these students vary greatly.

Programs for English Language Learners

School districts across the nation are aggressively considering how to address the academic needs of ELLs more effectively and more thoroughly. The challenge is to help students learn English, become literate, master content areas, and succeed academically. Although the greatest numbers of students classified as ELL remain at the elementary level, the greatest challenge is for students who arrive in middle and high school. The kinds of programs serving these students vary in the degree that native language is used and in the approach they take in teaching academic content (Linquanti 1999).

Bilingual models use the students' native language to teach academic subjects, such as science or social studies, reading, or language arts. These vary widely and can be transitional (early-exit), maintenance (late-exit), or two-way immersion programs. Bilingual classes are usually taught by teachers fluent in the native language of ELLs. In mainstream classes, native language can also be used as support with the help of bilingual paraprofessionals.

English programs referred to as *English as a Second Language* (ESL), *English Language Development* (ELD), or *English for Speakers of Other Languages* (ESOL) use instruction in English only to teach about the language, to communicate in English, and to develop language skills while preparing students to study grade-level content. The programs can operate by themselves or as part of bilingual programs in which they are one of the instructional components. Two widely used models are

Pull out. In this model, students are pulled out of some classes in order to receive English instruction. They typically receive all instruction in English. Students may receive as little as 20 minutes or as much as several hours each day. A major disadvantage of this option is that, while learning English, learners still must make up for the content instruction lost while they are away from the classroom.

Push in. The ELD teacher or aide provides support within the mainstream classroom and helps the classroom teacher adapt instruction as necessary. Inclusion allows for collaboration between language and content-area teachers. Thus, learners receive the benefit of more opportunities for interaction with speakers of English while avoiding the self-esteem and low motivation pitfalls usually attached to being sent out of the classroom.

Sheltered English programs address the language proficiency and the content-area components by using specially modified instruction that links language learning with academic work. Drawing from well-established best practices, sheltered English classes in science and other content areas provide opportunities for learners to improve their English while being able to keep up in academic areas that reflect corresponding grade-level curricula.

Although a wide range of programs are specifically designed to address the needs of ELL students, the reality is that many teachers of all grades find themselves in a regular self-contained classroom teaching either a subject or a variety of subjects to these diverse learners. Ever-increasing fiscal challenges and newly arrived students entering schools year-round result in ELLs often being mainstreamed into the regular classroom, ready or not. *Submersion* refers to mainstream classes in which no special language or content support is provided by trained specialists, and ELL students are forced to either sink or swim. In these situations, the regular classroom teacher is on his or her own constructing best practices to integrate the ELL student into the classroom.

A number of educators have suggested that the optimal program for ELL students is one in which ESL instruction is offered with mainstreaming, sheltered classes, and maintenance bilingual education (Richard-Amato 2003). Most educators do agree that ELLs need special support, that all teachers should have training in pertinent teaching strategies, and that programs should be designed and continually reassessed based upon the changing student population. In a

report on program alternatives for ELL students Genesee (1999, p. 4) says, "No single approach or program model works best in every situation. Many different approaches can be successful when implemented well. Local conditions, choices, and innovation are critical ingredients of success."

How a Second Language Is Learned

Maria sits in the first row of her fourth-grade mainstream class. She has barely spoken a word of English in class since her arrival in the United States two months earlier. She seems to understand directions, she listens and pays close attention to the teacher, and she seems to enjoy participating in class activities. But she almost never speaks.

Maria is an example of an English learner who is in the beginning stages of learning English. She is in the *silent stage* of language learning, which can last from a few hours to many months. She is learning English as she listens, but, as is typical of many beginning language learners, remains silent.

An understanding of how languages are learned can help mainstream teachers serve the culturally and linguistically diverse students in their classrooms. Current theories on how people learn languages are based on years of research in many fields, including education, linguistics, psychology, and sociology (Freeman and Freeman 2001).

Research in second-language acquisition suggests that there is a continuum of learning and that all learners progress through similar and predictable stages. The most commonly described stages are

- Stage 1: silent or preproduction stage,
- Stage 2: early production stage,
- Stage 3: speech emergence stage,
- Stage 4: intermediate language proficiency stage, and
- Stage 5: advanced language proficiency stage.

Beginning language learners typically begin with the silent stage, in which they may not speak but can understand words. They normally advance through stages during which they use a limited number of words and phrases, then make more complex statements. Reaching advanced proficiency typically takes students from five to seven years, at which time they know some academic language and can usually participate fully in grade-level activities. Understanding these stages can help teachers learn to accept a student's current stage and modify instruction to help the student progress to the next stage (Reed and Railsback 2003).

Language acquisition research also suggests that students learn English best when they are provided with language that is a little beyond their level of competence, which Krashen (1981) called *comprehensible input*. Students also learn when they are provided with opportunities to use language they have acquired through *comprehensible output* (Swain and Lapkin 1995). The opportunities for hearing English and using it to express oneself are critical to learning English. Not only do students need exposure to a language, but they also need to interact in meaningful, relevant contexts.

Many researchers have also noted language learning is best achieved in an anxi-ety-free environment. Krashen (1981) noted that an *affective filter* plays an important role in language learning. When students feel unthreatened and engaged in a classroom, their motivation is increased, and they are willing to do more risk-taking, which positively affects language learning.

Research has also provided much information on the different kinds of language students must learn. Cummins (1986) differentiated between social language—basic interpersonal communication skills (BICS)—and academic language—cognitive academic language proficiency (CALP). He showed the importance of cognitive demand and context in developing academic language of content instruction in the classroom. The average student takes two to five years to develop conversational skills, but four to seven years to develop academic fluency (Thomas and Collier 2002). Many studies suggest that students learn language best when they are focused on content-area studies rather than on grammar and the language itself (Freeman and Freeman 2001; Crandall 1987).

It follows that science academic language provides a highly suitable medium for promoting the development of the academic language proficiency ELLs are going to need if they are to be successful in school (Cummins 2000).

How Science Is Learned

Cassi has been outside for hours. When she is called in for lunch, she comes from the hill in the backyard with a plastic shoebox full of dirt, grass, and twigs. In the dirt is a plethora of wiggling and squirming worms.

Window
Into the
Classroom

Cassi's dad asked her what she was doing. "I am learning about worms," she replied.

Children are born into this world with a natural curiosity about the world around them. Cassi clearly illustrates this. Unfortunately, as children are conditioned by other forces such as home, school, and society, that natural curiosity is often taken away from them. A parent will say "Don't touch that; it is yucky," or a teacher will say, "I never liked science." Many of the people most important in the development of the inquiring child, even with the best intentions, undermine the natural learning of science. To counter this, educators and researchers in science education have moved toward a student-centered approach using an inquiry methodology while moving away from the more traditional content-centered approach characterized by an emphasis on memorizing facts (NRC 2000; Bybee 2002).

Research on exactly how children learn science dates from the work of John Dewey about 120 years ago and goes on to the insights provided by B. F. Skinner, Jean Piaget, and Jerome Bruner. Each of these great people added to the knowledge of and the presentation of how science is learned, from a systematic series of steps that must be followed to a knowledge developed from experience (Howe 2002).

The most popular theory on how science is learned stems from the constructivist theory of learning. Constructivists believe that students learn by fitting information together with what they already know and that learning is affected by context as well as students' beliefs and attitudes. Although the main philosophy of constructivism is gen-

erally credited to Jean Piaget (1896–1980), almost a century earlier Henrich Pestalozzi (1746–1827) "labeled rote learning as mindless, and he emphasized instead linking the curriculum to children's experiences in their homes and family lives" (Ornstein and Hunkins 1993, p.75). This is particularly relevant as we consider that science curriculum, teaching, and learning have been greatly influenced by policy and historical events, such as Sputnik in 1957. When the United States lost the race into space, then-President Eisenhower stated that no longer would the United States be second in math and science. The end result of this declaration was an efficiency curriculum with emphasis on covering more information (memorization) rather than on the process of doing science.

From Piaget, constructivist theory in science education broke into various branches (Novak 1977; von Glasersfeld 1989). One of these, contextual constructivism, is defined by how the learner interprets phenomena and internalizes these interpretations in terms of previous experience. This version of constructivism places significant emphasis on the role of culture (Soloman 1987; Millar 1989; Cobern 1991, 1993). It is this sort of concern that needs to be taken into account when addressing the academic needs of ELLs with their diverse linguistic and cultural perspectives.

Constructivism or a constructivist view puts the students, their interests, background, culture, and previous experiences and knowledge in a position of paramount importance in designing a curriculum that students can understand. One of the main methodologies of constructivism is the inquiry approach to teaching science, which involves use of tasks that integrate appropriate processes and information to construct meaningful learning experiences for students (Thier 2002). As explained in Chapter 7, a major proponent of this approach, Robert Karplus, even studied with Piaget in order to develop the learning cycle, the driving force of inquiry instruction in science education today (Fuller 2002).

Best Practices in Science Teaching and Language Teaching

There are many different opinions about which teaching methodologies constitute best practices in science teaching or in language teaching. Methodologies in both fields have changed over the years as discussed in Chapter 10. A comparison of current views on best practices suggests parallels between science and language teaching and also how to integrate methodologies of the two.

Science theories on learning influence current pedagogies. The constructivist theories described in the previous sections have broad implications for science teaching practices. One of the most reader-friendly works entailing best practices for a constructivist classroom is *A Case for the Constructivist Classroom*, written by Brooks and Brooks (1993). These authors provide five guiding principles of constructivism using

1. problems of relevance to students in instruction,
2. structured learning around primary concepts,
3. students' points of view,
4. adapted curriculum to address students' suppositions, and
5. assessments of students' learning in the context of teaching.

A second example of science instruction was set forth by Wheatley, who proposed that the teacher's role is to "provide stimulating and motivational experiences through negotiation and act as a guide in the building of personalized schema" (1991, p.14). Wheatley's problem-centered learning approach includes three components: tasks, groups, and sharing. He suggests that, in preparation for a class, teachers should select tasks with a high probability of being problematic for students. Second, the students should work on these tasks in small groups. During this time, the teacher attempts to convey collaborative work as a goal. Finally, the class is convened as a whole for a time of sharing. Wheatley notes that the teacher should then lead the class in a discussion in which each of the groups presents its solution methods, inventions, and insights. This opportunity for sharing promotes higher-level thinking and reasoning skills and often leads the students to further "conversations" and independent thinking.

Another pedagogical approach specifically related to science education is Saunders' four-step approach (1992). The first step is to organize hands-on investigative labs. These are problem centered and differ from the traditional "recipe" labs in that there are no or few prescribed methods or procedures. Students must use their own schema to formulate expectations about what is likely to be observed. The second component is active cognitive involvement. This is in contrast to the passive learning that takes place in many "teacher-centered" classrooms. Saunders explains that learning is made meaningful through activities like "thinking out loud, developing alternative explanations, interpreting data, participating in cognitive conflict (constructive arguing about phenomena under study), development of an alternative hypothesis, the design of further experiments to test alternative hypothesis, and the selection of plausible hypotheses from among completing explanations" (p.140). The third component involves students working in small groups. Saunders says that small-group work tends to stimulate a higher level of cognitive activity among larger numbers of students than does listening to lectures and thus provides expanded opportunities for cognitive restructuring. Fourth, he considers higher-level assessment. Although he does not fully address alternative assessment, the literature on it is vast and suggests that "considerable advances in learning follow when teachers use assessment formatively" (Harlen 2001, p. xi).

In their discussion of language teaching practices, Herrell and Jordan (2004) note that many of the approaches teachers use with all students can be used with English language learners with additional planning. Their premises for effective instruction are that teachers should

* provide instruction that ensures that students are given comprehensible input,
* provide opportunities to increase verbal interaction in classroom activities,
* use teaching strategies and grouping techniques that reduce the anxiety of students, and
* provide activities in the classroom that offer opportunities for active involvement of the students (p. 5).

Jameson has summarized four best practices for teachers of language learners to use in the classroom (1998):

1. Make content more understandable to students by providing nonverbal cues such as pictures, demonstrations, and hands-on learning.
2. Increase interaction by using cooperative learning and project-based learning.
3. Increase thinking/study skills by asking higher-order "thinking" questions such as "What would happen if ..." and having high expectations for all students.
4. Use a student's native language to increase comprehensibility when possible.

In recent years, with the focus on developing language and literacy skills for all children in academic subjects, best practices have focused on the teaching of language through content. Haley and Austin (2004) suggest that interactive learning activities with content and student collaboration in the classroom provide the means for helping students develop more complex academic language skills. Becijos (1997) outlines a number of specially-designed-academic-instruction-in-English (SDAIE) strategies for teachers to use, including

* creating instruction that relates to students' prior knowledge,
* tailoring teacher talk to students' English language proficiency levels,
* allowing students to process material in a variety of formats, and
* using assessment methods that allow students to display learning in a variety of ways.

Echevarria, Vogt, and Short have outlined a number of lesson components using the SIOP Model for content area and language instruction. These are described in Chapter 7 in the context of designing science lessons (2004).

We suggest that the following aspects of teaching relate to both science and language learning and, when applied to the classroom, can provide successful learning opportunities for all children:

1. Research-based learning models: Many science and language strategies stem from a constructivist philosophy. Both science and language are learned progressively, and there are a series of steps that build upon student investigative skills.

2. Cooperative learning: Instruction incorporating learning in pairs and groups allows for interaction between students, which is critical for ELL students. Each child works within a group but has individual accountability to the group.

3. Active cognitive involvement (hands-on—heads-on): Passive instructional models in science and language should be used sparingly. The more senses a student uses in learning, the more effective the learning is. Kinesthetic learning experiences with realia are particularly effective in both science and language.

4. Input from students: Teachers in both science and language need to understand the lived experience of children and adjust planning to accommodate the culture, experience, ability, learning styles, or interests of the child.

5. Student-centered classrooms: Students and student learning should be the focus of the classroom teaching. A teacher cannot be both "the sage on the stage and the guide on the side."

6. Integration of subject matter to convey connections to the experiential world: Children learn best when information is integrated and put into a real-world context. Learning content is not isolated from one subject to another.

7. Interaction, discussion, reflection, and teacher flexibility: The key here is for the teacher to be flexible in both curriculum and pedagogical strategies, to work at a pace that suits the majority of students, and to include discussions and reflection as part of every learning experience.

Although this may not be an exhaustive list, instructional practices that benefit English language learners in the science classroom benefit all students and must be continuously reevaluated as students' proficiency levels and needs change. The science teacher reading through principles and methodologies for teaching language may realize that he or she incorporates many of them in science instruction. Conversely, the language teacher reading through the best practices in science may see how similar they are to language teaching practices. This is the nexus of the book and the recurring theme throughout all of the chapters. None of this information is new. What is new is that the two fields are recognizing the similarities, identifying differences, and coming together to accommodate practices to address the academic needs of a linguistically diverse student population.

References

Anderson, S. 2004. The multiplier effect. *International Education* 13 (3): 14–21.

Becijos, J. 1997. *SDAIE: Strategies for teachers of English learners.* Bonita, CA: Torch Publications.

Brooks, J., and M. Brooks. 1993. *The case for a constructivist classroom.* Alexandria, VA: Association for Supervision and Curriculum Development.

Bybee, R., ed. 2002. *Learning science and the science of learning.* Arlington, VA: NSTA Press.

Cobern, W. 1991. Contextual constructivism: The impact of culture on the learning and teaching of science. Paper presented at the annual meeting of the National Association for Research in Science Teaching. Lake Geneva, WI. April 7-10.

Cobern, W. 1993. Contextual constructivism. In *The practice of constructivism in science education*, ed. K. Tobin, 51–69. Washington, DC: AAAS.

Cummins, J. 1986. Empowering minority students: A framework for intervention. *Harvard Education Review* 15: 18–36.

Cummins, J. 2000. *Language, power and pedagogy: Bilingual children in the crossfire.* Clevedon, England: Multilingual Matters.

Crandall, J., ed. 1987. *ESL in content-area instruction.* Englewood Cliffs, NJ: Prentice Hall Regents.

Echevarria, J., and A. Graves. 2003. *Sheltered content instruction: Teaching English-language learners with diverse abilities*, 2nd ed. Boston: Allyn and Bacon.

Echevarria, J., M. Vogt, and D. Short. 2004. *Making content comprehensible for English learners*, 2nd ed. Boston: Pearson.

Freeman, D., and Y. Freeman. 2000. *Teaching reading in multilingual classrooms.* Portsmouth, NH: Heinemann.

Freeman, D., and Y. Freeman. 2001. *Between worlds: Access to second language acquisition*, 2nd ed. Portsmouth, NH: Heinemann.

Fuller, R. 2002. *A love of discovery: Science education—the second career of Robert Karplus.* New York: Kluwer.

Genesee, F., ed. 1999. *Program alternatives for linguistically diverse students.* Educational Practice Report No. 1. Santa Cruz, CA, and Washington, DC: Center for Research on Education and Diversity.

Halley, M., and T. Austin. 2004. *Content-based second language teaching and learning: An interactive approach.* Boston: Allyn and Bacon.

Harlen, W. 2001. *Primary science: Taking the plunge*, 2nd ed. Portsmouth, NH: Heinemann.

Herrell, A., and M. Jordan. 2004. *Fifty strategies for teaching English language learners*, 2nd ed. Upper Saddle River, NJ: Pearson.

Howe, A. 2002. *Engaging children in science*, 3rd ed. Upper Saddle River, NJ: Pearson.

Jameson, J. 1998. Three principles for success: English language learners in mainstream content classes *(From Theory to Practice,* Issue No. 6). Tampa, FL: Center for Applied Linguistics, Region XIV Comprehensive Center. Retrieved December 12, 2002, from *www.cal.org/cc14/ttp6.htm.*

Kindler, A. 2002. *Survey of the states' limited English proficient students and available educational programs and services 2000–2001 summary report.* Washington, DC: NCELA.

Krashen, S. 1981. *Second language acquisition and second language learning.* New York: Pergamon Press.

Linquanti, R. 1999. *Fostering academic success for English language learners: What do we know?* San Francisco, CA: WestEd. Retrieved January 18, 2005, from *www.wested.org/policy/pubs/fostering.*

Millar, R., ed. 1989. *Doing science: Images of science in science education.* Philadelphia, PA: Falmer Press.

National Research Council. 2000. *How people learn: Brain, mind, experience, and school.* Washington, DC: National Academy Press.

National Science Teachers Association. 2000. Position Statement on Multiculturalism (Available online at *www.nsta.org/positionstatementandpsid=21).*

Novak, J. 1977. *A theory of education.* Ithaca, NY: Cornell University Press.

Ornstein, A., and F. Hunkins. 1993. *Curriculum: Foundation, principles, and issues*, 3rd ed. New York: Pearson.

Reed, B., and J. Railsback. 2003. *Strategies and resources for mainstream teachers of English language learners.* Portland, OR: Northwest Regional Educational Laboratory.

Richard-Amato, P. 2003. *Making it happen: Interaction in the second language classroom.* White Plains, NY: Longman.

Saunders, W. 1992. The constructivist perspective: Implications and teaching strategies for science. *School Science and Mathematics* 92 (3): 136–141.

Shin, H. 2003. *Language use and English-speaking ability: 2000.* Washington, DC: U.S. Census. Retrieved September 14, 2004, from *www.census.gov/prod/2003pubs/c2kbr-29.pdf.*

Siegel, H. 2002. Multiculturalism, universalism, and science education: In search of common ground. *Science Education* 86: 803–820.

Soloman, J. 1987. Social influences on the construction of pupil's understanding of science. *Studies in Science Education* 14: 63–82.

Swain, M., and S. Lapkin. 1995. Problems in output and the cognitive processes they generate: A step towards second language learning. *Applied Linguistics* 16 (3): 371–391.

TESOL (Teachers of English to Speakers of Other Languages). 1997. *ESL standards for pre-K-12 students.* Alexandria, VA: Author.

Thier, M. 2002. *The new science literacy: Using language skills to help students learn science.* Portsmouth, NH: Heinemann.

Thomas, W., and V. Collier. 2002. *School effectiveness for language minority students.* At *www.ncela.gwu.edu/pubs/resource/effectiveness.*

NCBE Resource Collection Series No. 9. Washington DC: National Clearinghouse for Bilingual Education. Retrieved January 17, 2005 from *www.ncela.gwu.edu/ncbepubs/resource/effectiveness.*

United States Census Bureau News. 2003. Press release CB03-157. Washington, DC: Author. Retrieved September 20, 2004, from *www.census.gov/Press-Releases/archives/census_2000/001406.html.*

United States Census Bureau News. 2003. Summary tables on language use and English ability: 2000 (PHC-T-20). Washington, DC: Author. Retrieved September 12, 2004 from *www.census.gov/population/www/cen2000/phc-t20.html.*

Wheatley, G. 1991. Constructivist perspectives on science and mathematics learning. *Science Education* 75 (1): 9–21.

von Glasersfeld, E. 1989. Cognition, construction of knowledge, and teaching. *Synthese* 80 (1): 121–140.

Further Reading

DeBoer, G. 1991. *A history of ideas in science education. Implications for practice.* New York: Teachers College Press. This is a foundational piece in science education covering topics since the early 1900s. The book discusses historical aspects of science education and the roots of modern-day practices.

Díaz-Rico, L.T. 2004. *Teaching English learners: Strategies and methods.* Boston: Allyn and Bacon. A comprehensible and critical discussion of both theory and methodology. The author addresses how teachers can tailor their instruction to varied instructional contexts that take into account ELLs' linguistic and cultural backgrounds. The discussion of standards-based learning is particularly relevant as it focuses on cultural and sociopolitical considerations. Very well organized and current.

Harlen, W. 2001. *Primary science: Taking the plunge,* 2nd ed. Portsmouth, NH: Heinemann. An updated revision of a long-time classic, this is a highly readable text that focuses on the role of the teacher in the primary school classroom. Intended for the generalist teacher, it is a clear and articulate examination of the rationale for and how to create meaningful learning experiences for science students. Drawing on the role of inquiry in helping students' understanding of scientific processes and skills, the author provides a step-by-step framework for effective science teaching. The book will appeal to ESL and science teachers alike.

Herrera, S., and K. Murry. 2005. *Mastering ESL and bilingual methods: Differentiated instruction for culturally and linguistically diverse (CLD) students.* Boston: Allyn and Bacon. This book examines research-based methods that are effective with diverse students and that promote linguistic and academic achievement. Thematic lessons, a highlight, include specific suggestions and practical applications.

Section II

Strategies for Planning, Teaching, Assessing, and Extending Learning

Planning Science and English Instruction: One Teacher's Experience

Anne Baumgarten and Marie Bacher

"Soil," said Mr. Alexander as he held up a cup of coarse soil.

"Sieve," he said as he held up one of the tools to be used for the day's soil explorations.

"Particles," Mr. Alexander said as he poured the cup of soil through the sifting tool. Then he gestured for his fourth-grade class to say the words with him as he repeated the demonstration of sifting particles of soil.

When he finished introducing the tools and materials needed for the day's investigations, he turned to a beginning English language learner (ELL) student and asked her to repeat the words, soil and sieve to him. Then he asked an intermediate ELL student for a short sentence describing what happened. Finally, he turned to an advanced ELL student and had a short conversation about where the student might have seen or used a tool such as a sieve. He had made sure that the ELLs were able to hear the language forms connected to the science learning in a way aimed at each specific fluency level.

Then Mr. Alexander had the students work in groups to recreate in their science notebooks what he had presented to the class. At one table, Kevin Kim, a student with basic communications skills in English repeated in Korean what Mr. Alexander said to his table partner See He Park, a beginning ELL. Kevin quickly helped her draw pictures of the tools in her science notebook and label them.

After this brief introduction, Mr. Alexander had one student from each group gather the materials needed to begin the investigation. The students followed the direction cards written in concise English with graphics supporting each step. Mr. Alexander began to roam the room, targeting students who he had already determined would need extra support to be successful throughout the lesson.

Window *Into the* Classroom

In this chapter, the process of planning to teach science to English language learners is outlined through the eyes of one fourth-grade classroom teacher, Mr. Alexander. The chapter follows Mr. Alexander's shifts over several years in thinking about teaching science and about supporting his ELLs in acquiring academic English language. He first thought of his instruction as based on his state's content standards. Then he began to think of the abilities and needs of his students as the driving force behind his instruction. Finally, he imagined his instruction as a dynamic interaction between standards and students.

Learning science and learning a second language are cognitive processes that support each other. The science process skills—which include observing, predicting, communicating, classifying, and measuring—are almost the same skills needed for learning language—which include seeking information, comparing, ordering, classifying, analyzing, inferring, justifying, solving problems, synthesizing, and evaluating. (More information on strategies for teaching ELLs process skills and language skills is in Chapter 4.)

This narrative demonstrates how, over the years, one classroom teacher realized that these cognitive tasks were the keys students needed to integrate content instruction with language acquisition. He shifted his emphasis from teaching bits of information to teaching the foundational thinking skills his students would need to understand scientific concepts and acquire a new language. By combining his own research and training on science inquiry and English language development (ELD), experimenting with instructional strategies, and then creating ways to scaffold his own thinking to support his instructional practice, he was able to integrate content and language acquisition instruction.

Using Standards to Plan Instruction

When Mr. Alexander started planning his science units and lessons as a new teacher, he often felt as if he were being asked to build a picture puzzle using pieces without color or defining shape and without any glossy photographs to refer to. The pieces of his puzzle included standards for teaching reading, writing, mathematics, science, social studies, physical education, and visual and performing arts. Most of the different content standards did not have any overarching relationship to each other. He felt that, if he could not easily connect the chunks of information covered in the standards, his students would have trouble making connections.

Along with the reading and writing standards, Mr. Alexander had to look at his district's essential English language development (ELD) standards for English language learners. These standards were based on state ELD standards, reviewed, and then focused to reflect his district's English language arts standards. This derivation from the state ELD standards was intended to help teachers understand the linguistic pathways that English language learners go through to achieve the language arts standards.

But the ELD standards added another confusing layer to the puzzle because they addressed multiple fluency levels—advanced, early advanced, intermediate, early intermediate, beginning—and grade levels—K–2 and 3–5—in contrast to the language arts standards that were English-language and grade-level specific (Buck 2000; Carr and Lagunoff 2003; Fathman, Kessler, and Quinn 1992; Short 1991). There was more to consider: Mr. Alexander had to include a host of teaching strategies such as choral reading (groups of students chorally present a poem or other reading selection); scaffolding (a teacher provides templates or patterns to support student understanding); total physical response (a teacher dramatizes vocabulary); RAFT (a teacher determines the role of the writer, audience, format, and topic so students can check for understanding); and co-op (students complete a project in teams).

These teaching strategies did not seem connected to the specific science content standards, to the language arts reading standards, or to the ELD standards. In trying to solve the problem of constructing a big picture of teaching content standards to English learners, Mr. Alexander often felt that the big picture was three-dimensional, rather than two-dimensional (see Figure 1).

This figure shows the connections among language arts, ELD standards, instructional strategies, and science content standards. The proficiency levels of language learners are indicated in the figure as beginning (B), early intermediate (EI), intermediate (I), early advanced (EA), and advanced (A).

Figure 1: Integrating standards and instructional strategies

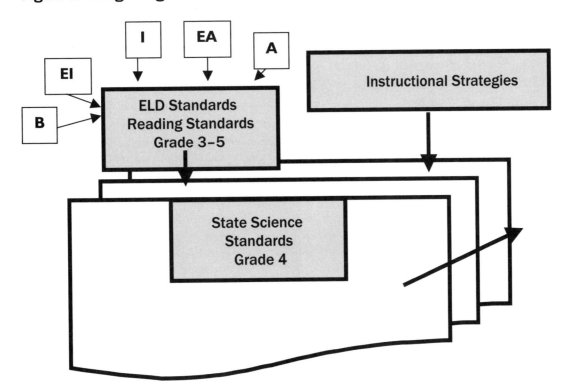

Planning to teach the soil unit his first year, Mr. Alexander began by looking at the state grade-four science standards. A science methodology class during his teaching induction program and his participation in a summer inquiry institute based on the work of scientists and designers at San Francisco's Exploratorium had made him a firm believer in inquiry-based science instruction. So he planned to set up five investigation stations in soil to engage the students. To understand the content fully, he read the curricular materials that covered the standards, including the Earth Full Option Science System (FOSS) Kit adopted by his district and the chapter on earth materials from his district's science textbook. He asked his grade level teammates for advice and read a few ideas on a teacher website. Finally he was ready to introduce the unit to his students. He set up the materials, handed science notebooks to his students, wrote directions on the board, and asked the students to get started.

At the end of the day, he reflected on his hard work. He had observed two groups of students who had been heavily engaged with the materials and who had recorded their questions and observations in their science notebooks. In another group of students, half the team worked with the activities while the other half quietly copied from their classmates. Another two groups of students either watched others or chatted. This was not the learning experience he had envisioned.

The next day he decided to try something different. As the students worked through their investigations, Mr. Alexander circled the room and asked questions to determine his students' understanding. Their answers revealed a wide range of conceptions and comfort in engaging in an open-ended exploration. He read through their written work and found a mix of abilities in students' expressing themselves clearly through scientific drawing and writing. Throughout that first year, Mr. Alexander tried a variety of teaching strategies—ranging from directed inquiry experiences to worksheet type science notebooks—to support his teaching. He felt he was working hard to teach to the standards but that not all his students were learning the concepts covered in his lessons.

Using Student Information to Plan Instruction

As he gained experience, Mr. Alexander began to realize that the make-up of his class was different from year to year. At times he had students who were fluent in their primary languages, which included Vietnamese, Cambodian, Japanese, Chinese, Cantonese, Tagalog, Punjabi, Korean, Spanish, Russian, Greek, Persian, and Illocano (a Philippine language). Some students told him they were not fluent in their primary language; in essence they had no language base to build their academic knowledge upon. These students were especially challenging when it came to content instruction.

Each student came to him with a complex knowledge of language and culture that affected his or her ability to learn and understand the science lessons. Many of his English learners were Spanish speakers, and much of the ELD research and curricular materials provided by his district were aimed at these students. Over the years, however, he discovered that not all of his Spanish-speaking students had similar background experiences to build upon when learning new ideas. In addition, students who came from Asia spoke languages that were very different from English in phonemic structure and in print-sound relationship. Now he felt that the big picture of teaching science to English learners must be at least a four- or five-dimensional problem.

Mr. Alexander found that, after using the state content standards to plan his units, he then had to gather information at the beginning of each school year on each student to see what adjustments he would need to make for all students to be successful. He would first meet with the ELD aide to get the previous school year's California English Language Development Test (CELDT) scores in listening and speaking, reading, and writing. Federal and California state law require that school districts give an English-language proficiency test to K–12 students whose home language is not English. The test must be given within 30 calendar days after a student enrolls for the first time in a California public school. Districts must also administer the CELDT annually to identify English learners until they are reclassified as fluent English proficient (FEP).

The aide reminded him that the composite score for the domains of listening, speaking, reading, and writing would not give him the depth of information he needed to support his students in learning science. The composite represented an idea of how a child had performed on a standardized assessment but not of the specific skill strengths and weaknesses in each language domain. Mr. Alexander asked the aide to show him the breakdown of each domain to get a complete picture of his ELLs' language proficiency.

He learned that ensuring access for all students in the science curriculum is an enormous job. Taking the time to gather information on every English learner was crucial to each student's success. Teachers need to collect that information to decide what cognitive tasks, process skills, and academic language forms to highlight during scientific discourse (Avalos and Lee 2002).

To maintain accurate records for each ELL in his class, Mr. Alexander developed a language arts performance chart using the table function in Microsoft Word to track English language proficiency (ELP) scores and literacy development scores from the previous year. A typical chart had scores from the state's language arts standardized test, including word analysis, reading comprehension, literary response, written conventions, and writing strategy. The chart also had scores from the school district's literacy performance-based assessment (PBA) and a section for a student's home language, cumulative file information, and other anecdotal notes. Mr. Alexander could now analyze the students and develop lessons to maximize student academic success and achievement. (See Table 1 as an example of a record-keeping chart.)

Mr. Alexander kept records on all his students but found that, when he focused in depth on two students in his class, all his students benefited. After gathering information on all his students, he chose two whose language needs might jeopardize their conceptual learning in science. First he chose Tu Vo, a Vietnamese girl who had come to the United States in the past year, to be

Table 1: Language arts performance chart

Student and Home Lang	LEP Level/ CELDT scores	District Reading PBA	State Reading Comp	State Reading Response	State Word Analysis and Vocab	State Writing Strategies	State Writing Conven	Notes: CUM Anecdotal Lang Arts Portfolio
Mai, Oravy Cam- Not fluent	1	Text Level 20	230	200	200	210	215	
Vo, Tu\n\nViet-\nFluent	NA	Text level 0	NA	NA	NA	NA	NA	Outgoing Parent Support

one of his target students. Her cumulative file said that she entered school proficient in her primary language. Adjusting to the U.S. classroom had been very difficult for Tu because the cultural norms for student behavior and classroom atmosphere were very different in Vietnam. She did not understand group work or the movement of students to different learning centers, such as the classroom library and computers. Tu's schooling in Vietnam had given her academic language and content knowledge in her first language, however, allowing her to draw on this background to develop concepts in the science content area. Mr. Alexander realized that she would go through both a cultural and linguistic transition period before her prior academic knowledge could support her new learning.

Because Tu was a beginning English language learner, Mr. Alexander decided the language objectives for the soil lesson would be action and scientific equipment words: *pouring*, *sieve*, and *magnifying glass*. Knowing that peer tutoring is an effective strategy, he

decided to put Tu in the same group with Chris Nguyen, a U.S.-born Vietnamese-speaking student. Mr. Alexander made sure Chris knew it was all right to help and speak to Tu in Vietnamese during class.

The other student Mr. Alexander targeted that year was Oravy Mai, a student new to the school. His scores indicated he was an intermediate English language learner and could comprehend complex speech. But he still needed a fair amount of repetition of academic vocabulary and language functions such as comparing and contrasting. His cumulative file said that Oravy spoke only his native language, Cambodian, when he started school. His last school had offered an ESL pull-out program, but his instruction had been all in English. Now that he was in the fourth grade, he had lost his ability to speak his native language and it was hard for him to understand his parents and grandparents who mainly spoke Cambodian. Oravy had never developed literacy in his primary language. Mr. Alexander noticed two other students who spoke more than one language were also at the intermediate proficiency level. After reviewing his district's English development standards, he decided the intermediately proficient students' language objectives would be labeling pictures of the sieves and soil with short sentences describing the shape, color, and size of the soil particles.

By looking into the literacy and cultural background information of his students, Mr. Alexander became more aware of the abilities and limitations of different English language learners. Through understanding the academic and cultural background of each student, Mr. Alexander was able to create a menu that tiered the most appropriate cognitive task and academic language objective to each student's fluency level. He knew that, with close monitoring and effective feedback, English language learners could be supported. Mr. Alexander supported his own instructional practice by carrying a clipboard on which he recorded the lesson objectives and his anecdotal notes on student understanding.

Putting All the Pieces Together

Mr. Alexander's instructional practice described in the "Window Into the Classroom" section reflects his learning to build connections for himself among all the different standards he was required to use. He mentally mapped connections between the science standards, the language arts standards, and the ELD standards. He planned the lesson described above by focusing on his state's fourth-grade science standards on the properties of soils and the listening and speaking standards for beginning ELLs. The science concepts he

intended to teach included the properties of soil, such as color and texture, capacity to retain water, and ability to support the growth of many kinds of plants. Each investigation he planned included highlighting the basic science process skills as well as the integrated process skills, including controlling variables, making models, and relating. The language skills he decided to emphasize included the language function of describing and the language forms of nouns, pronouns, adjectives, and prepositional phrases.

Over the years, Mr. Alexander had found ways to see the big picture of teaching English through science. Somewhere in this learning process, he realized that instead of seeing different bits and pieces of his curriculum, or his teaching methods, or student abilities, he was seeing a whole, unified picture that was made up of various interconnecting pieces. His picture of his science and language arts curriculum became like the jigsaw puzzle box shown in Figure 2, a picture to guide him in constructing the instructional activities that would ensure every one of his students had access to science concepts while developing English language proficiency.

As described in the "Window Into the Classroom," when Mr. Alexander started his unit planning for this year, he reflected on how much he had grown professionally in being able to fit all these pieces together. The steps he followed were
- review the essential content knowledge,
- review process skills, and
- map these in a table.

Table 2 shows how content knowledge and process skills can be mapped.

Figure 2: The pieces of the puzzle

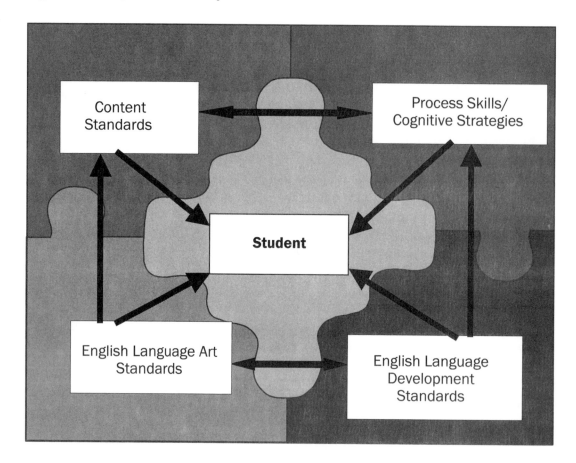

Mr. Alexander then reviewed the state English language arts standards to determine what language knowledge and skills the students would be learning at the same time they would be learning the science concepts. Finally, he reviewed his district ELD standards to determine how to support the variety of English learners he knew he would have in his class.

Part of his thinking breakthrough occurred from research material in Dutro and Moran's (2003) *Rethinking English Language Instruction: An Architectural Approach*, on breaking up the meaning of the acronym CALP (cognitive academic language proficiency) (Cummins 1986) into three areas (see Table 3).

Dutro and Moran propose that language functions are the purposes and uses of language. They drive learners to connect thoughts and language and help the brain create a framework for communicating concepts and principles of specific content areas.

Mr. Alexander had a thinking shift when he realized that these were the scientific process skills of questioning, predicting, hypothesizing, concluding, and analyzing,

Table 2: Fourth-grade content standards and science process skill benchmarks

Earth Materials Rocks and Minerals	Observing	Communicating	Planning and Conducting
Characteristic of minerals can be used to identify a mineral. Minerals are building blocks of rocks. Rocks are formed under different conditions. There are three classifications of rocks. The rock cycle: One type of rock can change into another. Earth changes over time. Soils have properties of color, texture, and the capacity to retain water and support growth of plants.	Differentiate observation from inference (interpretation). Know that scientists' explanations come partly from what they observe and partly from how they interpret their observations. With guidance, classify objects based on appropriate criteria. Measure and estimate weight, length, or volume of objects.	Draw accurate scientific illustrations with proper labeling. Construct and interpret charts and graphs from measurements. Write a procedure that someone else could follow.	With guidance develop a testable question. Formulate and justify predictions. Make and follow a basic plan to conduct multiple trials to test one variable. Identify necessary data to collect in an investigation. With guidance, interpret data and draw conclusions that answer the question being tested.

Table 3: Three areas of cognitive language proficiency

C	AL	P
Cognitive Tasks or Functions of Language	Academic Language or Language Forms	Proficiency
explain, infer, analyze, draw conclusions, synthesize, compare/contrast, persuade	syntax or sentence structure, grammatical features, and word usage	ease of comprehension and production, automaticity in reading and writing, and appropriateness of discourse

making him realize that science and language were intimately linked. To map all the various pieces of the puzzle, Mr. Alexander used a table to track all the pieces he needed to consider when planning instruction (see Table 4).

Table 4: Planning tool for "Separating Soil Particles"

Content Knowledge	Process Skills/ Cognitive Tasks	Language Skills Functions and Forms	Instructional Strategies to Support ELs
Different soils have different properties	**Observation** Differentiate observation from inference Classify objects based on appropriate criteria Measure objects	**Functions:** Compare/Contrast **Forms:** Vocabulary	Venn Diagram Pair/Share Total physical response Word bank
	Communicating Draw and label materials and procedures	**Functions:** Compare/Contrast **Forms:** Vocabulary	
	Planning and conducting investigations Questioning Predicting Explaining	**Functions:** Procedural writing Labeling	Chart on procedural writing elements with icons

Conclusion

Mr. Alexander knew that a lot of his learning about putting the pieces of the instructional puzzle together stemmed from being able to build concept bridges from one teaching pedagogy to another. He found that, when introduced to a new educational method or idea such as the Biological Sciences Curriculum Study's 5E learning cycle of science (engage, explore, explain, elaborate, and evaluate), he would keep on using it. Similarly, when introduced to the language arts pedagogy of *handover of responsibility*, in which the reader assumes responsibility for applying smart reading behavior in order to gain and maintain understanding through strategic teacher feedback (Routman 2002),

Mr. Alexander would cling to this methodology in his language arts block. When he attended district professional training on supporting ELLs by using sentence frames developed for each proficiency level, he felt another new idea had been added to all the other ideas he had been exposed to over the years.

Mr. Alexander found that, when he could map mentally the connections between science and language arts, he could then teach his students to map connections mentally. He knew that the best practices in science teaching involve facilitation, collaborative group work, and a limited, judicious use of information giving. He also knew that language is acquired through social interactions that are engaging, meaningful and purposeful. The connections for him were creating a safe learning community in his classroom, getting to know his students through their eyes and not his own, and teaching foundational cognitive skills. To remind himself that these beliefs would help him work smarter, not harder, he posted the following quote from R. Caine and G. Caine (1994) in his classroom:

"Knowledge becomes natural when it is sufficiently connected with what else is already known. These patterns of interconnectedness are what we call 'maps.' To help students create sophisticated maps in the brain, teachers must not present subject matter in isolated, meaningless pieces. Rather the student needs to experience a sense of wholeness."

References

Avalos, M. A., and O. Lee. 2002. Promoting science instruction and assessment for English language learners. *Electronic Journal of Science Education* 7 (3).

Bay Area School for Excellence in Education (BAS-EE). 2003. Science resources for teachers, 2003. Available online at *http://sustainability.terc.edu/index.cfm/resources*.

Buck, G. A. 2000. Teaching science to English-as-a-second-language learners. *Science and Children* 67 (9): 38–41.

Caine, R., and G. Caine. 1994. *Making connections: Teaching and the human brain*. Reading, MA: Addison Wesley.

Carr, J., and R. Lagunoff. 2003. *Map of standards for English learners*. San Francisco: WestEd.

Cummins, J. 1986. Empowering minority students: A framework for intervention. *Harvard Education Review* 15: 18–36.

Dutro, S., and C. Moran. 2003. Rethinking English language instruction: An architectural approach. In *English learners: Reaching the highest level of English literacy*, ed. G. G. García, 227–258. Newark, DE: International Reading Association.

Fathman, A., C. Kessler, and M. E. Quinn. 1992. Teaching science to English learners, grades 4–8. Retrieved December 9, 2004, from NCELA Program Information Guide at *www.ncela.gwu.edu/pubs/pigs/pig11.htm* .

Routman, R. 2002. *Reading essentials: The specifics you need to teach reading well*. Portsmouth, NH: Heinemann.

Short, D. 1991. Integrating language and content instruction: Strategies and techniques. Retrieved December 9, 2004, from NCELA Program Information Guide at *www.ncela.gwu.edu/pubs/pigs/pig7.htm*.

Further Reading

Avalos, M. A., and O. Lee. 2002. Promoting science instruction and assessment for English language Learners. *Electronic Journal of Science Education* 7 (3). This paper addresses the issues of science instruction and assessment with ELLs. The authors concentrate on the importance of science learning for all students and the current status of science instruction and assessment for ELLs. They review and describe effective policies and practices for science instruction and assessment for ELLs. The paper concludes that participation in science instruction can promote literacy development and English language proficiency for all students.

Jarrett, D. 1999. *Teaching mathematics and science to English-language learners*. Portland, OR: Northwest Regional Educational Laboratory. This publication is part of a series, "It's Just Good Teaching." This publication gives educators research-based instructional strategies with real-life examples from classrooms in the Northwest. It includes informative sections on second language strategies with content instruction, collaborating with other teachers, and involving the family. The scenes from the classroom are authentic and provide a window into the classroom.

Peregoy, S., and O. Boyle. 2005. *Reading, writing, and learning in ESL: A resource book for K–12 teachers,* 4th ed. Boston: Allyn and Bacon. This is a comprehensive methods text for teaching ELL children in the many different settings in school. Complete with background information, instructional strategies, and learning scenarios, the book guides the reader through the complicated issues and tasks of adapting instruction for diverse learners in diverse learning environments.

Chapter 4

Strategies for Teaching Science to English Learners

Deborah Maatta, Fred Dobb, and Karen Ostlund

Ms. Antal, a teacher in a middle school sheltered science class, begins her lesson by showing her students a plant and asking them, "How can I kill this plant?"

"Put it in the trash," one student says.

Ms. Antal puts the plant in the trash and asks, "How will this kill the plant?"

"No light," another student says.

"Oh, plants need light," Ms. Antal says and writes the information on the board. "What else can we do to kill the plant?"

"Don't give water," a student replies.

Ms. Antal writes, "Plants need water."

After a number of students have given their responses, Ms. Antal explains that the amount of water and light and the temperature are physical conditions and tells her class they will do an investigation to see how physical conditions affect population growth. Using an overhead projector, Ms. Antal goes through each step of the investigation, carefully modeling what the students will need to do. She has simplified the language of the lab report and added sentence frames. The students work with their teams and set up two terrariums, carefully making sure conditions are the same in each terrarium except for varying one physical condition to see its effect on population growth.

Observations are recorded over time, and conclusions are then generated from the observations and data. A rich class discussion comparing and contrasting the students' findings summarizes the investigations.

Window *Into the* Classroom

Teachers who teach science play a key role with English language learners (ELLs). The science classroom provides a language-rich environment where ELL students can express their understandings of the world. Inquiry-based science instruction as depicted in the "Window Into the Classroom" has the potential to involve ELLs in observing, recording, and communicating their experiences in a direct and immediate way. Regardless of changing federal, state, and local policies and practices regarding the education of ELLs, science instruction has a distinct place in the development of academic language. For no reason should ELLs be excluded or denied science instruction because of their limited proficiency in English. In fact, students who become proficient in the language of science are likely to become enthusiastic academic achievers in other areas of study. This chapter is guided by two fundamental principles:

- Inquiry science provides shared experiences. In other words, it offers an arena in which ELLs can try out their maturing ideas about scientific phenomena using their expanding second-language skills. What is learned in science through English remains as part of one's understanding of the universe and represents a step in one's growth into a second language. Thus, the student who plans, plants, observes, and records growth of his or her terrarium in English brings purpose and significance to both life science and English language development (ELD) skills.

- There is a correspondence between learning increasingly complex scientific content through the scientific processes and incremental demands for an ever-expanding vocabulary and literacy skills as ELLs progress through the levels of English language proficiency in speaking, listening, reading, and writing. In the science classroom, students learn to use contextually appropriate language that is accurate, precise, and objective.

An emerging body of research, particularly in science, has substantiated the benefits of inquiry-based instruction for students who are developing proficiency in English (Amaral, Garrison, and Klentschy 2002; Valadez 2002; Gibbons 2003). Most research targets what might be called the English language learner-friendly hallmarks of the inquiry classroom (Stoddart 2000). Among the chief characteristics of inquiry-based instruction are common experiences for students, hands-on activities, links to prior knowledge, student collaboration, and opportunities to read, listen, talk, and write about events and experiences. All of these characteristics enhance and give depth to the ELL experience in science. In addition to practicing inquiry-based science and focusing on science process skills, there are several strategies teachers can use to help English language learners learn science. These strategies relate to

- connecting with students,
- teacher talk,
- student talk,
- academic vocabulary,
- reading skills,
- writing skills,
- collaborative learning,
- scientific language, and
- process skills of inquiry.

When these strategies are used in combination with inquiry science, the ELL student gains a better understanding of content science and English (Dobb 2004). This chapter will discuss each of these strategies and give examples of how they may be used in K–8 science instruction.

Connecting With Students

How can I connect with English learners in my classroom? Not all issues in delivering effective programs for English language learners involve standards, pedagogy, and teacher content knowledge. For many ELLs, success in science comes only once they see a place for themselves within the science-related professions, or see role models in action, or receive encouragement from their teachers. Unquantifiable, the feelings of belonging and acceptance are vital for students to envision their own success. The following statement by an English learner captures a sense of purposefulness in learning science that comes from positive experiences:

> I am trying to learn more English every day in school, but I don't want to lose my culture and Spanish language. The things I want to fight for, and the things that no one will take away from me, are the following: becoming a medical doctor, maintaining my language and culture, and to be able to work together with people who speak other languages, or who come from a different cultural background than mine.... I will reach my dream of being a doctor.... I will be a bilingual medical doctor and I hope I will see you in my office some day.
>
> (Seventh-grade student, Azusa Unified School District, California)

Often overlooked in quality science teaching is the affective domain, essential in making a positive connection with English learners. While trying to receive, interpret, and explain science content, English learners frequently ask themselves a series of questions in our classrooms. Teachers will likely find some resonance in these questions. Here are a few examples:

- Does this teacher know who I am?
- Does this teacher care about me?
- Does this teacher want me to succeed?
- Does this teacher realize that I am not intellectually limited, even though I am not able to express myself completely in English?
- Does this teacher understand the fear of ridicule and embarrassment I must overcome every time I open my mouth to speak, participate in a group, or hand in written work?
- Does this teacher see me as a potential contributor to scientific knowledge?

These are very relevant concerns of students who need to be assured and reassured that their teachers are aware of the issues behind each question and are willing to reach out to each student.

Teachers who are able to connect to ELL students are successful in communicating that students are not alone in their struggle to participate, that support is available, and that former English learners have succeeded in the science program. Equally important on the part of the instructors is the recognition that the English learner population is diverse. Instructors must avoid making false, defeating assumptions about how demanding science content knowledge and academic English are and the career goals and the futures of English learners (Dobb 2004).

There are some simple ways to connect with the linguistically and culturally diverse students in the classroom. A welcome sign written in the languages spoken by students in the class helps students and families to feel welcomed and recognized (see Figure 1).

You, as a teacher, can label science equipment in English and in the home language of your students. Make an effort to find real world examples that most students can relate to. For example, in a nutrition lesson, use a food such as chicken soup—which can be found in most cultures—rather than pizza, hamburgers, or hot dogs—which are typically American. You can give examples of scientists and researchers who are former English language learners or from diverse backgrounds. The best way to connect with linguistically and culturally diverse students is to learn more about the cultures of the students so that you can draw their home experiences into the classroom.

Teacher Talk

How can I help English learners understand what I say? Teacher talk needs to be at a linguistic level appropriate for English language learners. Making teacher talk understandable is one of the keys to ensuring full participation of English learners in a science classroom. How teachers adapt their speech and presentations depends on the proficiency levels of the students in the class—beginners will need more adaptations than intermediate level students.

Using simple sentence structures that are familiar to students, slowing your rate of speech, repeating and rephrasing main ideas, and enunciating clearly will help English learners understand. For instance, when explaining that diffusion is the "movement of molecules from regions where they are more concentrated to regions where they are less concentrated," you can also say "diffusion is when molecules move from one place where there are a lot of molecules to a place where there are few molecules."

Teachers can also ask students to summarize key points to be sure they understand. Support classroom talk with visual cues such as diagrams, pictures, gestures, dramatization, realia, modeling, and other visual aids that provide contextual clues and meaning. You should also do comprehension checks. For example, show a drawing of diffusion and a drawing of sand falling to the bottom

Figure 1: Welcome poster example

of a cup of water and ask students to explain which is diffusion and why; or ask students to model diffusion in the classroom with each student playing the role of a molecule.

When explaining academic tasks, present the tasks step-by-step and provide visual support such as a demonstration or picture of what students are expected to do. Provide written as well as oral instructions for students. When presenting a hands-on activity for students, model the activity without giving away the results. Simplify lab reports, and, when possible, add drawings and picture cues to provide an additional source of information (see Figure 2).

Student Talk

Should students speak another language in class? Primary language support is appropriate for English language learners at various levels of proficiency development, especially for ELLs who have not reached an intermediate level of English language proficiency and do not have a firm literacy base. Research shows that primary language support facilitates cognitive and academic development (Hakuta and August 1998).

You need not be bilingual to promote primary language support in the classroom. For students with strong literacy skills in their native languages, provide bilingual dictionaries and other-language editions of the textbook and tests. Group students with a common primary language to facilitate primary language support. Students can discuss a concept in their native language first and then, together or individually, express their thoughts in English. You can ask students to translate words in their native language. Use these words to create a bilingual science word list. Sharing the words with the class can help English-only students become more familiar with the sound of a new language.

Draw attention to similarities among the vocabularies of Romance and Germanic languages and English by examining these words. English learners with developed primary language proficiency in a Romance or Germanic language bring to the science classroom one frequently untapped advantage: the use of cognates. Students need to learn how to recognize shared root words and cognates in English and their own languages. For example, there are an estimated 10,000 to 15,000 Spanish-English cognates. Just examining one commonly used English word, *bug*, provides an example of the potential vocabulary richness available to students through examining cognates. The scientific word for bug is *insect*, which shares the Latin root *insectum* with the Spanish cognate *insecto*.

Recognize that not all students have a strong scientific vocabulary in their primary languages. It is more likely that the students will know the primary language equivalents for most vocabulary used in everyday conversation but not for science-related words. For example, if the third-grade ELL is learning about electricity, chances are he or she is learning about a circuit for the first time and does not have the primary language vocabulary to accompany this knowledge area. Words such as *circuit, transformer, watts, energy efficiency,* and *insulator* are probably new to the student in both the home language and English. In such cases, present the vocabulary in English and reinforce it without using direct translation in the home language.

Figure 2: Adapted lab report

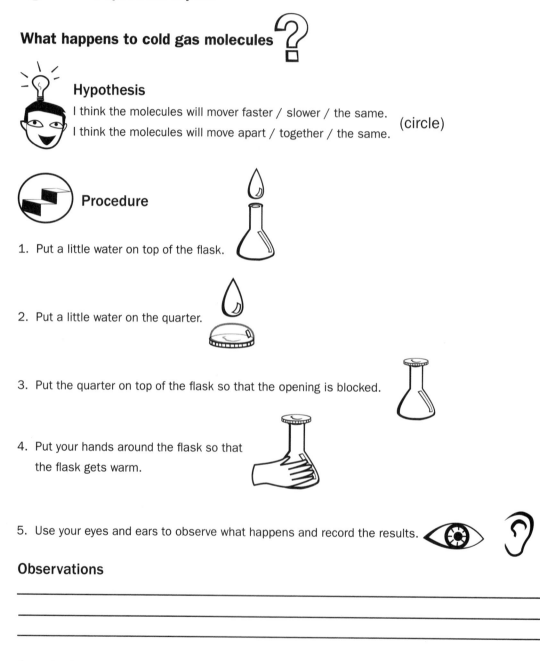

What happens to cold gas molecules?

Hypothesis

I think the molecules will mover faster / slower / the same.

I think the molecules will move apart / together / the same.   (circle)

Procedure

1. Put a little water on top of the flask.

2. Put a little water on the quarter.

3. Put the quarter on top of the flask so that the opening is blocked.

4. Put your hands around the flask so that the flask gets warm.

5. Use your eyes and ears to observe what happens and record the results.

Observations

_____

_____

_____

Conclusion (tell what happened to the molecules)

The quarter moved because _____

_____

In the classroom that affirms linguistic diversity, teachers encourage ELLs to expand their primary language literacy skills by connecting with personal experience, and by consulting classmates, instructors, parents, dictionaries, and primary language science materials. This affirms the value of the student and his or her language, deepens student understanding of new vocabulary, and strengthens literacy.

Academic Vocabulary

What are effective ways to teach vocabulary to English learners? In *Accelerating Academic English: A Focus on the English Learner*, Scarcella (2003) advises that vocabulary-building activities target not just meaning, but also such related issues as parts of speech, fre-

quency, appropriateness in scientific writing, which words are commonly used together, and pronunciation. Derived from Blachowicz and Fisher (2002), there are four main principles that should guide vocabulary instruction:

1. Students should be active in developing their understanding of words and ways to learn them. This can include semantic mapping, word sorts, and illustrating vocabulary words. During semantic mapping (Hanf 1971), students generate a list of words related to a topic. Students then categorize the words and create a semantic mapping of words related to the topic as illustrated in an activity on classification using the five kingdom classification system (see Figure 3). During a word sort (Bear et al. 2000), the

Figure 3: Semantic map for the five kingdoms

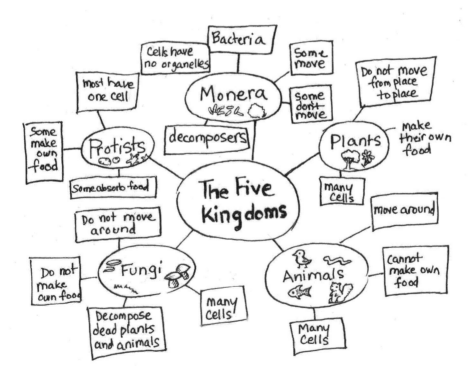

teacher provides a list of words related to the topic and students categorize them according to meaning, similarity in structure, derivations, or sounds. For example, words related to the five senses—such as *ears, eyes, nose, tongue, taste, touch, smell, fingers, skin, see,* and *hear*—could be sorted and an appropriate label created for each group (for example, *Body Part* or *Action*).

2. Students should personalize word learning through activities such as using mnemonic strategies, creating personal dictionaries, and creating games to review words.

3. Students should be immersed in words. The classroom should be a language-rich environment that focuses on words and learning words. Word walls (see Figure 4), science dictionaries, and focusing on root words and comparing and contrasting words with similar roots—for example, *hydrogen, dehydrated, hydroelectric*—will help students learn and use vocabulary.

4. Students should be provided with multiple sources of information to learn words. Key vocabulary should be taught and reviewed more than once. Multiple sources of meaning should be provided. Realia, drawings, pictures, video, gestures, and dramatization are examples of ways to provide sources of meaning other than the definition of the word.

Because a basic core of approximately 2,000 high-frequency words accounts for most words in academic writing, using one of the many well-researched, high-frequency word lists available is highly recommended. General academic language as well as science discipline-specific vocabulary is likely to pose challenges for most ELLs, so it is advisable to consult both a general academic language list as well as a science-specific list. The following are representative high-frequency word lists:

*   Corson, D. 1997. The learning and use of academic English words. *Language Learning* 47 (4): 671–718.
*   Coxhead, A. 2000. A new academic word list. *TESOL Quarterly.* 34 (2).
*   Taylor, S. E. 1989. EDL core vocabularies in reading, mathematics, science and social studies. Austin, Texas: Steck-Vaughn/EDL. 428.2.

It is important to teach vocabulary within the scientific context, not in isolation, and to avoid asking students to memorize vocabulary lists without content support.

Reading Skills

How can I help my ELLs better understand their reading assignments? Linking to prior knowledge is a key to deeper comprehension of print. Linking to prior knowledge helps students activate their schema—knowledge frameworks stored in memory—for a given topic. Schema theory (Vacca and Vacca 2005) holds that learners draw on their schemata to make sense of new information by seeing how it fits with what they already know.

When working with culturally and linguistically diverse students, recognize that their experiences may differ greatly from those of English-only students and strive to understand where gaps in prior knowledge may exist. Where gaps exist, use classroom experiences to help broaden a student's knowledge of a subject. For example, students can visit a construction site where machines are used and discuss their obser-

Figure 4: Example of word wall for a solar system lesson

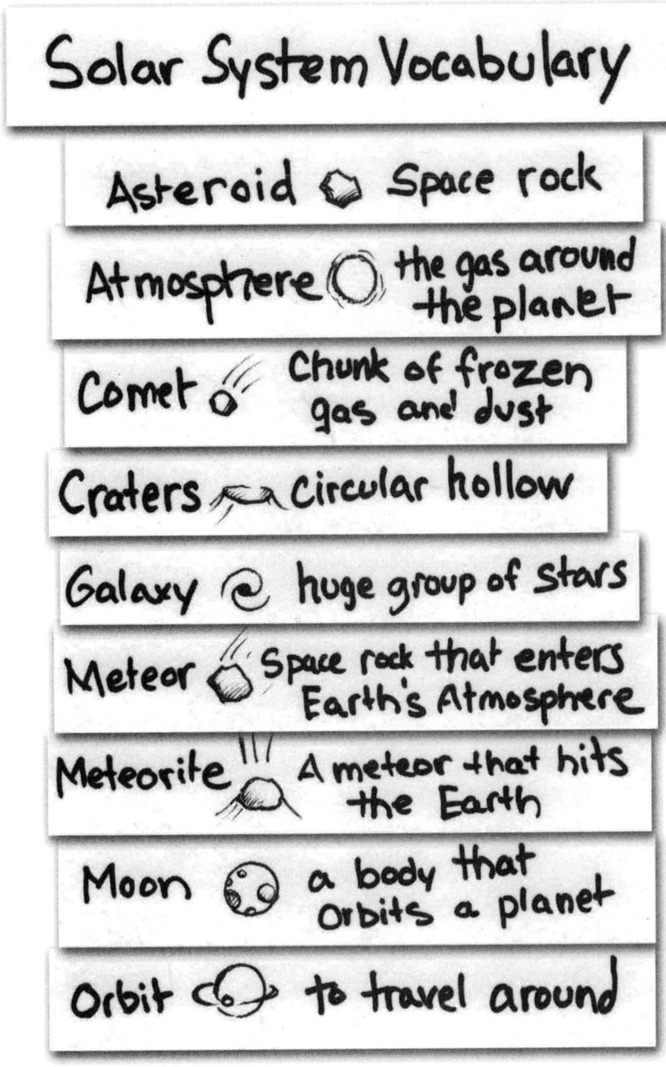

Solar System Vocabulary

Asteroid — Space rock

Atmosphere — the gas around the planet

Comet — Chunk of frozen gas and dust

Craters — circular hollow

Galaxy — huge group of stars

Meteor — Space rock that enters Earth's Atmosphere

Meteorite — A meteor that hits the Earth

Moon — a body that orbits a planet

Orbit — to travel around

the familiar K-W-L chart, which records what students know (K), want to know (W) and have learned (L) (Ogle 1986). To better adapt the K-W-L for a science lesson, use the THC strategy. The THC strategy, similar to the K-W-L, asks three questions: what do I **T**hink, **H**ow am I going to find out, and what do I **C**onclude (Crowther and Cannon 2004). The THC strategy simulates scientific thinking, promotes thinking about the nature of science, relates to science process skills, and helps you assess students' prior knowledge.

Effective reading comprehension activities focus on important habits such as previewing material, recognizing chapter headings, identifying introductions, reading every first sentence in a paragraph, understanding visuals and graphs, summarizing, and answering end-of-chapter questions. English language learners in contact with unfamiliar science print material need to have information enhanced, expanded on, and made clear through

vations in the classroom. They can view a video on the ocean or visit an aquarium before reading about oceans. Use systematic ways of acknowledging student prior experience and knowledge and sharing it with others. One way to access prior knowledge and teach scientific thinking skills is to use re-presentation of text. Text re-presentation encourages students to probe beneath the surface of the text for deeper understanding. Ways of re-presenting a text include role-playing, cooperative dialog writing, and genre-transforming exercises (Donahue et al. 2003). In role-playing, students act out the text they have read. For example, students might read about the life cycle of the butterfly and then reenact the cycle complete with narration read from the text or rewritten by the students. In cooperative dialog writing, students work with a team to write a dialog based on a text. For example, after reading about a scientific laboratory team working on the development of a life-saving drug, students imagine what the team members might discuss among themselves and then write the dialog as a team. In genre-transforming exercises, students rewrite a passage using a different genre. For example, students examine a laboratory report and then create sequenced charts representing each step in the procedure described.

You can also use visual aids, demonstrations, graphic organizers, and textbook adaptations to make the textbook content comprehensible. Hands-on activities and collaborative learning experiences provide rich experiences from which English learners can derive meaning and understanding of scientific concepts. Provide these types of activities before students are introduced to the textbook or lecture explanation of the concept.

Writing Skills

What can you do when your English language learners don't express themselves well in writing? Because academic writing poses a challenge for most English language learners, you should scaffold written assignments to give them the support

Figure 5: Student choice sentence frame

? Does the mass change during a chemical reaction?

Hypothesis: I think the mass will (circle one):

increase

decrease

stay the same

because_____

they need to complete the assignment. As they gain skills, students perform at higher levels independently. One of the keys for working with ELLs who are improving their writing skills is to focus on one or two specific writing objectives for each assignment. Any language learner can be utterly discouraged by turning in an assignment only to have it returned covered in red ink. Instead of correcting every error a student makes, correct only errors that impede understanding and errors that pertain to the targeted objectives.

Adapt written assignments so that English language learners can more easily provide their answers in English (Bravo and Garcia 2004). Provide sentence frames to assist students in phrasing their answers. Instead of asking ELLs to write out their hypothesis, provide three logical choices and ask students to circle one and write why they chose that answer (see Figure 5, "Student choice sentence frame"). Discuss the hypothesis and the reasoning behind each hypothesis first, and then highlight the needed vocabulary before students are asked to write.

Graphic organizers such as outlines, charts, Venn diagrams, and concept maps can help English learners organize their writing and provide the needed vocabulary and language for a written assignment. Use the data table (Table 1) to record observations on plant cells (elodea) and animal cells to help prepare students for writing assignments on comparing and contrasting plant and animal cells.

Students with very limited English proficiency show their understanding in a variety of ways. ELL students can demonstrate their knowledge through visual representations. For example, ask them to: "draw and label the parts of a volcano," "arrange chairs in the classroom to reflect the solar system," and "draw and illustrate a word web." Figure 6 is an example of how an ELL student drew and labeled the relationships of predator, prey, and producers in a food web.

Collaborative Learning

How can I help English learners to participate fully in group work? Cooperative learning experiences—in which students have a group goal and success at achieving this goal depends on the individual learning of

Table 1: Plant and animal cell comparison

Type of Cell	Nucleus?	Cell Wall?	Cell Membrane?	Other parts?
Animal				
Elodea Cell (plant)				

Figure 6: Predator-prey-producer relationship map

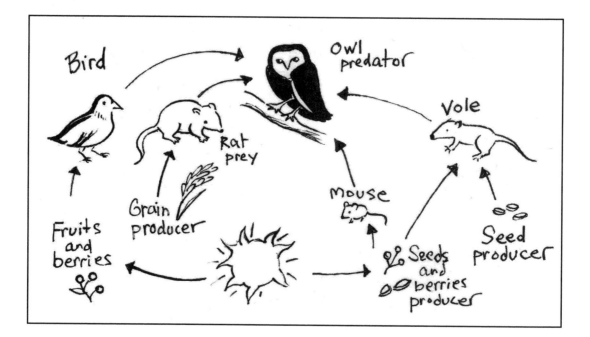

all group members—have proved to increase student achievement (Slavin 1987). Cooperative learning activities (Johnson and Johnson 1989) stress that students should be interdependent and held to individual accountability.

For example, when asking students to pair up for a task such as discussing the results of an experiment, use a timer so each student speaks for one minute, then ask students to write a two-sentence summary of the experiment result in their journals. Equal participation is ensured because each student speaks for one minute, and all students are accountable because they will individually write in their journals. During a group discussion, students should take turns recording the group ideas. Use different-colored pens to show that each group member has written a response.

When planning and implementing col-laborative learning experiences, clearly articulate the roles and expectations for all students' participation in the activities. For example, during a hands-on investigation students should have these roles:

- recorder, who records the observations from the group;
- reporter, who explains the groups' observations to the class;
- illustrator, who draws and explains the phenomena observed; and
- spokesperson, who is allowed to talk to the teacher during the investigation (Johnson and Johnson 1989).

Positive interdependence is structured successfully when group members perceive that they are linked together in such a way that one cannot succeed unless everyone succeeds. You must design and communicate group goals and tasks in such ways that make students believe they sink or swim together. In structuring positive interdependence, highlight that

1. each group member's efforts are required and indispensable for group success, and
2. each group member has a unique contribution to make to the joint effort because of his or her resources and/or role and task responsibilities.

This reinforces each student's commitment to the success of group members as well as to his or her own success and is the heart of cooperative learning. If there is no positive interdependence, there is no cooperation (Johnson and Johnson 1989).

Scientific Language

How can I help students use scientific language? A hands-on science classroom can be a rich, highly motivating language laboratory. One of the keys to building on ELL experiences and leading ELLs to reading and writing about science is engaging them in purposeful, guided instructional conversations about their science experiences and investigations.

In *Mediating Language Learning: Teacher Interactions With ESL Students in a Content-Based Classroom,* Pauline Gibbons (2003) found that student growth results when teachers assist ELLs "in moving from registers expressing their firsthand experience in oral language to those expressing academic knowledge in writing." In the science classroom, this continuum might move from a student's talking in a small group about an experiment on solids, liquids, and gases; to telling the teacher what he or she had learned about solids, liquids, and gases; to writing about solids, liquids, and gases in a journal; and finally to writing a short report comparing solids, liquids, and gases. First guide students to use scientific language during their oral explanations of their learning. During these instructional conversations, signal a need for clarification of ELL talk and provide clues for modifying the language of the ELL students. Model and focus on key vocabulary and grammatical structures. Instead of restating what a student means to say, ask questions to elicit more information. For example, Gibbons' research uncovered this conversation between a teacher and students about an investigation with magnets:

Teacher: Tell us what happened.

Student 1: Um, we put three magnets together. It still wouldn't hold the gold nail.

Teacher: Can you explain that again?

Student 1: We, we tried to put three magnets together to hold the gold nail. Even though we had three magnets, it wouldn't stick.

Teacher: Tell us what you found out.

Student 2: We found out that the south and the south don't like to stick together.

Teacher: Now let's start using our scientific language.

Student 2: The north and the north repelled each other and the south and the south also repelled each other but when we put the two magnets in a different way they attracted each other.

The teacher elicits clarification from students as they stretch their linguistic resources to communicate their learning. ELLs take responsibility for making themselves understood to teachers who are interested in developing effective speakers who can describe, explain, question, hypothesize, and persuade.

Familiarity with scientific discourse patterns is crucial in helping English language learners understand and use scientific language. Students need to practice reading, listening to the material, and using discourse patterns common to science. Figure 7 includes some of those common patterns.

One way to introduce the patterns is to use children's trade books on scientific topics such as weather, environmental concerns, or the water cycle. A list of outstanding science trade books is featured annually in the March issues of the National Science Teachers Association journals *Science and Children, Science Scope,* and *The Science Teacher.* Additionally, a web resource for a comprehensive list of alternative written materials for science is found at the California Department of Education website at *www.cde.ca.gov/ci/sc/ll.* These sources may be used to enrich scientific language.

Students can be guided into restating and expressing written information using scientific discourse patterns. For example, after reading and discussing a book about how pollution can ruin a river and how people can save it, use the discourse pattern labeled "make predictions" (I think _____ will _____ ____.) to prepare students for answering the question, "What do you think will happen to the river?" Students' answers will vary and may include responses such as "I think the river will remain a clean river," or "I think people will get sick from swimming in the river." Completed patterns reveal a wealth of topic-specific vocabulary and diverse ways of answering questions and explaining the text. Student responses should be written down and shared with all. As students read and re-read the completed sentence patterns, they develop confidence and fluency with the new structures. Eventually, with repeated practice, students will become comfortable with these patterns and add them to their talk about science and other subjects.

Figure 7: Examples of sentence patterns in scientific discourse

DESCRIPTION:	The _____ has _____ and _____.
CITE INFORMATION:	Here we see that_____.
ESTIMATE:	Looking at the _____, I think there are _____.
RETELL:	First, _____next, _____ and then,_____.
MAKE PREDICTIONS:	I think _____will_____.
GIVE AND SUPPORT OPINIONS:	I think_____is_____because_____.
CAUSE AND EFFECT:	The_____had_____,so_____.
DRAW CONCLUSIONS:	The_____ is _____because _____.
HYPOTHESIZE:	If_____had_____, then_____would have_____.
PERSUADE:	As we just saw in the experiment, _____ does _____ because_____.

Science content is enriched and language enhanced when you provide students with opportunities to include common scientific functions in their work. If students are to carry out observations, you must be conscious of what language functions are involved and work at practicing and refining those functions through student use. Functions are not inherently beginner, intermediate, or advanced. For example, although *description* is commonly cited as a basic function, describing is done with varying degrees of detail, accuracy, and appropriate science vocabulary. There are differences among saying that an apple is green, describing the exact thickness of its skin, or citing its water content. All are ways of *describing* the object, but reflect varying degrees of linguistic precision and scientific knowledge. Coach ELLs steadily to approximate the full and accurate use of carrying out those functions throughout their stages of competence in English. The following three activities combine science content with communicating information:

Prediction—Present a picture sequence of the butterfly's life cycle, and leave out the last part of the sequence. Ask students to draw the missing part.

Explaining and Analyzing—Present students with a table showing growth of plants under different conditions and ask them to analyze and explain the data.

Distinguishing fact from opinion—Have students read an editorial about water con-servation and ask them to make a T-chart displaying fact and opinion contained in the passage.

Scientific language functions or process skills are used whenever student speak, hear, read, write, or think as they structure sensory input from their environment. As they attempt to understand the world in which they live, they become proficient in using these skills and functions, making these tools the most powerful they have for producing and arranging information. Students may reach plateaus on which newly acquired process skills become integrated, used, and made functional. As they develop process skills, they pass through ways of thinking, each representing a different organization of experience, information, and knowledge, and each leading to a very different view of the world. (See Figure 8.)

Process Skills of Inquiry

How can teachers help English language learners develop basic science process skills of inquiry? Science as a process of inquiry guides students to appreciate the dynamic nature of science. Hands-on and minds-on

Figure 8: Examples of common scientific language functions (Dobb 2004)

Analyze	Distinguish fact	Predict
Calculate	from opinion	Provide evidence
Classify	Estimate	Question
Confirm	Identify	Reflect upon
Contrast	Interpret	Report
Defend a position	Justify	State
Describe	Observe	
Discuss	Persuade	

experiences are fundamental to learning science by inquiry (Ostlund 1992). Teaching science as inquiry involves a shift from depending on the textbook as the basic source of information to using the textbook as a reference. Laboratory activities are central as students investigate and inquire about the world and their own observations become the authoritative source of data. Students discover the facts, concepts, and laws of science in much the same way as the original discoveries were made rather than through the rote memorization of facts. The emphasis on firsthand observations in learning science reflects the belief that students should follow the processes used by scientists in learning facts and concepts (NRC 1996).

The process skills of science inquiry that students use to carry out hands-on and minds-on investigations include observing, communicating, classifying, measuring, predicting, inferring, formulating hypotheses, controlling variables, experimenting, defining operationally, formulating models, interpreting data, relating, and applying (see Figure 9).

These process skills were introduced in *Science—A Process Approach (SAPA)* using the insights of the psychologist Robert Gagne as a guide in the program design (AAAS 1964).

The science process skills listed above are very similar to the skills used in second-language acquisition. In many cases, a skill in science has either a similar or different name in language, both meaning basically the same thing (Herrell 2000). A teacher who helps students develop the science process skills of inquiry is helping them develop reading and language arts process skills at the same time. The processes of science inquiry are part of and central to other disciplines. A science teacher can make the content more accessible by providing opportunities for English language learners to engage in hands-on/minds-on activities using the process skills of inquiry. This provides a language-rich environment in which ELL students can develop their spoken and written English. Figure 10 indicates how the science process skills are related to English language acquisition skills.

Because English language learners come from such varying backgrounds, teachers who work with ELLs may need to take steps to build the skills and knowledge necessary for performing inquiry-based activities. Just as scaffolding provides the structure and support needed to construct a building, scaffolded inquiry provides essential support as students construct the skills and knowledge needed to build science literacy. As students progress through the stages of inquiry, the support provided by the teacher diminishes, and student ownership of the investigative process increases. This developmental process is essential for students to reach the ultimate goal of conducting science investigations independently, engaging in full inquiry.

The continuum of inquiry (Ostlund 2005), mentioned in literature since the late 1960s, is a series of developmental stages. Students progress through these stages to learn the skills and knowledge necessary to engage in inquiry (Jarrett 1997). The general sequence of the stages in the continuum of inquiry is outlined below.

Directed Inquiry—The beginning stage, directed inquiry, is teacher- or materials-directed. It provides a structured model of the inquiry process. Without this type of support and guidance, students cannot progress to asking and answering scientific questions independently (Cody 1998).

Directed inquiry introduces students to the essential features of inquiry and helps students reflect on the characteristics of the processes in which they are engaged.

Directed inquiry provides the foundation upon which subsequent stages of inquiry are built. This instruction is designed to provide students with the experiences necessary to learn and practice the processes of inquiry.

Guided Inquiry—In the second stage, guided inquiry, the teacher moves from the role of director to facilitator. Students continue to refine their inquiry skills based on the foundation developed during the directed inquiry explorations.

During guided inquiry, students have the opportunity to practice skills of inquiry

Figure 9: Basic process skills of science

Basic science process skills of inquiry for primary grades

Observing—the use of the five senses to gather data about objects and events

Communicating—the use of the spoken and written words, graphs, drawings, and diagrams to share information and ideas with others

Classifying—the use of observations to determine similarities and differences in objects and events to group or order objects or events into categories based on properties or criteria

Measuring—using both standard and nonstandard measures or estimates to describe the dimensions of an object or event

Predicting—stating the outcome of a future event based on a pattern of evidence

Inferring—the use of a logical thought process to show a relationship between observations or provide an explanation of an observation

Integrated science process skills of inquiry for intermediate grades

Formulating Hypotheses—a testable statement about the relationship between the manipulated and responding variables that can be tested experimentally

Controlling Variables—identifying and controlling variables in order to determine their effect on the outcome of an experiment

Experimenting—hypothesizing, designing an experiment to test the hypothesis, controlling variables, interpreting the data collected, and drawing conclusions

Defining Operationally—a definition framed in terms of your experiences

Making Models—developing a conceptual or physical representation of an object or event

Interpreting Data—analyzing and synthesizing data in order to draw a conclusion

Relating—the use of a logical thought process to determine the relationships involving interactions, dependencies, and cause-and-effect between and among objects and events

Applying—the use of a logical thought process to put scientific knowledge to use

Figure 10: Science process skills related to English-language acquisition

Basic Science Inquiry Process Skills	Using English to obtain, process, construct, and provide subject matter information in spoken and written form
Observing—the use of the five senses to gather data about objects and events	Following directions given orally Scanning the reading for new vocabulary Gathering information orally and in writing Defining and using target vocabulary in written communication Discriminating shapes, sounds, syllables, and word accents when reading Breaking words into syllables Pronouncing new words aloud
Communicating—the use of the spoken and written words, graphs, drawings, and diagrams to share information and ideas with others	Demonstrating verbal communication Developing and using new vocabulary through reading Persuading, arguing, negotiating, evaluating, and justifying Creating a variety of written communication Using target vocabulary to describe everyday objects and events Sequencing ideas Describing clearly Listing discoveries through a time line Describing chronological events
Classifying—the use of observations to determine similarities and differences in objects and events to group or order objects or events into categories based on properties or criteria	Comparing and contrasting information or characteristics Categorizing target vocabulary Arranging ideas Ordering and sequencing information Constructing charts that compare and contrast data
Measuring—using both standard and nonstandard measures or estimates to describe the dimensions of an object or event	Using skills to calculate and measure Selecting, connecting, and explaining information Using math skills to calculate and measure
Predicting—stating the outcome of a future event based on a pattern of evidence	Predicting the outcomes Generalizing Critically analyzing
Inferring—the use of a logical thought process to show a relationship between observations or provide an explanation of an observation	Distinguishing between fact and opinion Analyzing, synthesizing, and inferring from information Using information in other situations Identifying main ideas

with greater independence. Students are encouraged to think about variables, and they learn to plan for all the variables that may affect the outcome of an investigation.

Guided inquiry focuses student attention on learning particular science concepts. Students build science literacy and improve confidence in their abilities to do inquiry.

Full Inquiry—Full inquiry is one of the ultimate goals of science literacy. To conduct full inquiry, students must be able to apply the skills and knowledge developed in the previous stages of the continuum of inquiry.

According to the National Research Council (2000), full inquiry takes place if the following essential features of inquiry are present:

*   Questions are scientifically oriented,
*   Learners use evidence to evaluate explanations,
*   The explanations answer the questions,
*   Alternative explanations are compared and evaluated, and
*   Explanations are communicated and justified.

When inquiry is scaffolded, students will be able to succeed in reaching the goal of conducting independent inquiry. Without the developmentally appropriate framework provided by scaffolded inquiry, an unrealistic burden rests on the teacher (Colburn 2004). Moving from teacher-directed to teacher-facilitated to student-directed inquiries allows for a continual deepening of understanding of the skills and knowledge fundamental to conducting inquiry (see Figure 11).

When working with limited-English ELLs, it is most appropriate to begin their experiences at the level of directed inquiry, depending on the experiences they have had in the science classroom. As students gain both science process and English language skills, they will be able to perform student-directed inquiry activities.

Conclusion

The science classroom can be an exciting science- and language-learning laboratory for English language learners. The hands-on science classroom provides a rich environment in which ELLs can develop a scientific understanding of their world as well as gain the English skills necessary to communicate their discoveries and learning. The key to working with ELL students is adapting presentations, activities, and materials so that students with limited English proficiency can participate fully in the science classroom.

Try placing yourself in your English language learner students' shoes and ask yourself the question, "If I needed to learn this information in a language I don't understand well, what would I need?" Imagine how strategies such as direct vocabulary instruction, a simplified text, watching the teacher model the science experiment, and the chance to discuss the lesson with a peer who is proficient in the target language would help you participate in the lesson and feel more comfortable doing so.

As teachers of ELLs, we need to adapt our lessons to the language proficiency levels of our students to ensure that they are able to participate fully in the science classroom. In science, students can develop English language proficiency while learning the process skills and the content of science.

Figure 11: The continuum of inquiry

Processes of Inquiry	Beginning Stage Directed Inquiry	Transitional Stage Guided Inquiry	Final Stage Full Inquiry
Question	Students use question provided by the teacher, materials, or some other source.	Students are guided to refine and clarify questions developed with input from the teacher, materials, or some other source.	Students investigate a question that can be answered by doing descriptive investigations or experiments.
Predict or Formulate Hypotheses	Students are given a prediction for conducting a descriptive investigation or a hypothesis for conducting an experiment.	Students are guided to make a prediction for descriptive investigations or construct a hypothesis for experiments and revise predictions and/or hypotheses if necessary.	Students develop logical/reasonable predictions and/or hypotheses.
Investigate	Students are given the procedures and materials to conduct an investigation (descriptive investigation or experiment).	Students are given suggestions for procedures and materials that could be used to conduct an investigation (descriptive investigation or experiment).	Students devise a plan that takes all the variables into account and conduct an investigation (descriptive investigation or experiment).
Collect and Organize Data	Students follow step-by-step procedures provided by the teacher, materials, or some other source to collect and organize data into tables, graphs, and/or charts.	Students are given instructions to collect data, and tables, graphs, and/or charts are recommended to organize the data collected.	Students decide how and what data to collect and construct tables, graphs, and/or charts to organize the data collected.

(continued on next page)

(continued from preceding page)

Processes of Inquiry	Beginning Stage Directed Inquiry	Transitional Stage Guided Inquiry	Final Stage Full Inquiry
Analyze and Draw Conclusions	Students are given instructions on how to analyze the data, and guided to draw a conclusion to answer the question being investigated.	Students are given suggestions on how to analyze the data on their own and draw conclusions to answer the question being investigated.	Students determine what evidence is needed, analyze the collected data, and draw conclusions to answer the question being investigated.
Formulate Explanations	Students take synthesized data and are given step-by-step directions to formulate their explanation.	Students are guided in the process of formulating explanations from the data they collected, analyzed, and synthesized.	Students formulate their explanation after analyzing and synthesizing the data they collected.
Propose Scientific Explanations	Students are given scientific explanations.	Students are guided to reliable sources of scientific explanations and asked to compare this information to their explanations and make any necessary revisions.	Students independently examine scientific explanations from reliable sources and use the information to revise and strengthen their explanations.
Communicate Findings	Students are given step-by-step procedures for communicating findings and justifying a provided explanation with evidence from the investigation.	Students are given guidelines for communicating findings and justifying their explanations with evidence from their investigation.	Students use logical reasoning to communicate findings and justify their explanations with evidence from their investigation.

From Ostlund, K. L. 1992. *Science process skills: Assessing hands-on student performance.*
Menlo Park, CA: Addison Wesley.

Science teachers can make the content more accessible and at the same time provide a language-rich environment in which ELL students can develop their English as they use the process skills of inquiry.

References

Amaral, O., L. Garrison, and M. Klentschy. 2002. Helping English learners increase achievement through inquiry based science instruction. *Bilingual Research Journal* 26 (2). Available online at: *www.ncela.gwu.edu/miscpubs/nabe/brjlv26.htm.*

American Association for the Advancement of Science (AAAS). 1964. *Science—A process approach (SAPA)*. Washington, DC: Author.

Bear, D. R., M. Invernizzi, S. Templeton, and F. Johnston. 2000. *Words their way: Word study for phonics, vocabulary, and spelling instruction*, 2nd ed. Upper Saddle River, NJ: Merrill-Prentice-Hall.

Blachowicz, L. Z., and P. Fisher. 2002. Vocabulary instruction. In eds. R. L. Kamil, P. B. Mosenthal, P. D. Pearson, and R. Barr, *Handbook of reading research*, Vol. 3. (503-523) Mahwah, NJ: Lawrence Erlbaum.

Bravo, M., and E. Garcia. 2004. *Learning to write like scientists: English language learners' science inquiry and writing understandings in responsive learning contexts*. Paper presented at AERA Annual Meeting, San Diego, CA.

Cody, A. 1998. *Student questions: Foundations for inquiry*. Master's Thesis to Faculty of the College of Education, San Jose State University.

Colburn, A. 2004. Inquiring scientists want to know. *Educational Leadership* 62 (1).

Crowther, D., and J. Cannon. 2004. Strategy makeover: From KWL to THC: A popular reading strategy gets a science makeover. *Science and Children* 42 (1): 42–44.

Dobb, F. 2004. *Essential elements of science instruction for English learners*, 2nd ed. Los Angeles, CA: California Science Project.

Donahue, D., K. Evans, and T. Galguera. 2005. *Rethinking preparation for content area teaching: The reading apprenticeship approach*. Hoboken, NJ: WestEd and Jossey-Bass.

Gibbons, P. 2003. Mediating language learning: Teacher interactions with ESL students in a content-based classroom. *TESOL Quarterly* 37 (2).

Hakuta, K., and D. August, eds. 1998. *Educating language minority children*. Washington DC: National Academy Press:

Hanf, M. P. 1971. Mapping: A technique for translating reading into thinking. *Journal of Reading* 14.

Herrell, A. 2000. *Fifty strategies for teaching English language learners*. Upper Saddle River, NJ: Prentice Hall.

Jarrett, D. 1997. *Inquiry strategies for science and mathematics learning: It's just good teaching*. Portland, OR: Northwest Regional Educational Laboratory.

Johnson, D. W., and R. T. Johnson. 1989. *Cooperation and competition: Theory and research*. Edina, MN: Interaction Book.

National Research Council. 1996. *National Science Education Standards*. Washington, DC: National Academy Press.

Ogle, D. 1986. K-W-L: A teaching model that develops active reading of expository text. *The Reading Teacher* 39 (6).

Ostlund, K. 2005. Scaffolded inquiry. *CESI Science* 38 (1).

Ostlund, K. L. 1992. *Science process skills: Assessing hands-on student performance*. Menlo Park, CA: Addison-Wesley.

Scarcella, R. 2003. *Accelerating academic English: A focus on the English learner*. Oakland, CA: Regents of the University of California.

Slavin, R. E. 1987. Cooperative learning and the cooperative school. *Educational Leadership* (45).

Stoddart, T. 2000. *Integrating language, literacy and*

science instruction for English language learners: An annotated bibliography. Santa Cruz, CA: UC Accord.

Vacca, R. T., and S. L. Vacca. 2005. *Content area reading: Literacy and learning across the curriculum*, 7th ed. Boston: Allyn and Bacon.

Valadez, J. 2002. Dispelling the myth: Is there an effect of inquiry-based science teaching on standardized reading scores? Paper presented at the Second Annual Conference on Sustainability of Systemic Reform. Available online at *http://sustainability2002.terc.edu/invoke.cfm/page/729*.

Further Reading

Dobb, F. 2004. *Essential elements of effective science instruction for English learners*, 2nd ed. Los Angeles, CA: California Science Project. Examines the special challenges to science teachers working with English learners and the proposition that science teachers can be highly effective language teachers. Each element combines to produce an overall view of sheltered science instruction. Includes extensive resources, a Sheltered Instruction Observation Protocol, and a graphic organizer for instructional planning called "Targeting Academic Language Development for English Learners Through Scientific Investigation and Experimentation."

Herrell, A., and M. Jordan. 2003. *Fifty strategies for teaching English language learners*, 2nd ed. Upper Saddle River, NJ: Pearson. This practical, hands-on book provides 50 carefully chosen strategies to help ELL pupils understand content materials while perfecting their skills at speaking, reading, writing, and listening in English. Each strategy is accompanied by a definition, a rationale, and step-by-step implementation instructions, and all are specifically tied to the most current ELL standards.

NSTA Publications: The National Science Teachers Association (NSTA) publishes four journals available to science educators. Available online at *www.nsta.org*.

- *Science and Children* is the professional journal for the K–6 classroom teacher and also for many college and university level science methods instructors.

- *Science Scope* is the professional journal for the middle level teacher. The journal offers professional articles and activities relevant to teaching at the middle level.

- *The Science Teacher* is the professional journal for high school science teachers. The journal offers professional articles and activities that relate to teaching high school science.

- *The Journal of College Science Teaching* is the professional journal for college and university level instructors responsible for teaching content science as well as science methods. Articles and activities are relevant to content science instruction and methodology.

Chapter 5

Strategies for Assessing Science and Language Learning

Anne Katz and Joanne K. Olson

The fourth- and fifth-grade science class takes place in a classroom on the main floor of the elementary school, an older stucco building in one of the city's residential neighborhoods. Although reductions have resulted in fewer students in K–3 classes, this room barely holds the 29 gathered for science instruction. ELL students, most of who come from Chinese-speaking families, make up 28% of the school's student body. Mrs. Jackson, the science teacher, has completed English language development course work and passed the language development specialist exam in addition to her state teaching license. She draws on her science background and her understanding of the needs of ELL students to craft lessons and assessment for the diverse learners in her classroom.

Mrs. Jackson has designed a geology unit consistent with her state's fourth-grade science standards, focusing on environments of rock formation. She uses a guided inquiry approach, so students begin their exploration with activities such as examining various rock samples, taking measurements, and using dichotomous keys. They will then compare their observations to global geologic maps and determine where major rock types are most likely to be found. Mrs. Jackson will introduce earthquakes and volcanoes, locating these results of geological processes on a map of Earth's crust. As part of the unit, she has planned a field trip to a local natural science museum and an overnight trip to nearby mountains where her class will study not only geology but also the environment and the human history of the area.

She ensures that both instruction and assessment are multifaceted to engage students in many ways. To check her students' understanding of science concepts, she uses a range of assessments. She includes paper-and-pencil tests that include multiple-choice and short-answer questions that tap students' knowledge of science content. She also uses student performances so that students can show their understanding in other ways—for example, making a cross-sectional model of a volcano. Both kinds of assessment provide a window on students' grasp of how rocks are formed, the main concept for this unit.

To ensure that her ELL students can demonstrate their knowledge and skills, she uses approaches that are less dependent on language proficiency. For example, Ming Yee, who has low English proficiency, shows her understanding of volcanic processes by drawing a picture and providing a title and labels. As students like Ming Yee acquire more language proficiency, Mrs. Jackson incorporates more language-dependent assessments, such as sentence completion tasks, into her plan.

Her students maintain science notebook folders into which they put their notes, observations, reflections, and questions, and other assignments. These collections of student work serve as an ongoing assessment with several benefits:

- *They allow Mrs. Jackson to monitor when students have completed an assignment;*
- *They provide a visual display of the progression of student work over time; and*
- *They include student reflections on "I wonder" and "I think" statements—helping Mrs. Jackson and the students to monitor their speculations and developing conclusions over time.*

For ELL students, the folders provide multiple opportunities—across both time and kind of assessment—to show their understanding of science concepts and self-monitor their learning. The science folder also blends Mrs. Jackson's assessment agenda with her overall aim of guiding students into participation as a community of scientists who engage in doing science and learning from each experience, regardless of the outcome. The folders are one part of a larger assessment plan that involves performances, projects, written tests, and anecdotal records.

Learning is a cognitively active process of making sense of new information in light of what is already known. Mrs. Jackson, the teacher in the "Window Into the Classroom," is aware that two fundamental characteristics of the learning process, transfer and language dependence, frame our understanding of critical issues in teaching and assessing English learners in the science classroom.

Although educators frequently think of transfer in terms of a learner's ability to apply new knowledge to another context, we must also recognize that fundamental to learning anything new is the transfer of prior information to what is being learned (Bransford et al. 2000). The work of Vygotsky (1978) and others who study how people learn has provided valuable insight into the role of language in influencing learning. The way we conceptualize information is influenced by our language—that is, our very concepts are constrained by the words we use to define and describe those concepts. Therefore, when learning, people transfer relevant knowledge to the task at hand, and, using language, piece together a new understanding. This raises important issues for teachers of ELL students.

Issue 1: ELL students may struggle to learn because relevant prior knowledge is embedded in a different language and culture. ELLs may have elaborate concepts, but those concepts may be expressed in a different language—a language that may not have identical meaning when translated into English. And learning new concepts is far more than an issue of translation from one language to another. The very way that learners attend to information is partly dependent on cultural factors (Solano-Flores and Nelson-Barber 2001). For instance, how to solve problems, respond to a teacher's questions, and make observations are all shaped by assumptions shared by members of a culture. As a result, the learner (1) may not be able to wrestle with the underlying concepts and their fit or lack of fit with existing ideas or assumptions, and (2) may misinterpret new information because of inaccurate prior knowledge, language differences, and cultural differences.

Issue 2: Science is inherently language intensive. Many opportunities exist to engage students in hands-on experiences with the phenomena being studied. But expressing an understanding of the science concepts requires the use of language, and science is particularly language intensive. A quick examination of most science textbooks reveals that the amount of terminology rivals that of a foreign language textbook. For a language learner in school, the entire experience is a foreign language course, and the science terminology is a substantial additional burden. Reform documents such as the *National Science Education Standards* (NRC 1996) and international studies of student achievement (Schmidt et al. 1999) argue for a greater conceptual emphasis and a reduced emphasis on terms and definitions. Even when teachers introduce terms only when students need them, language in the science classroom is still difficult for language learners. Understanding a life cycle, for example, requires the learner to have a concept of life, its stages, and their order, and each of these ideas is expressed and understood with language.

This backdrop of the role of transfer and language in the learning process raises important issues about assessing language learners in the classroom. Researchers have asserted that meaningful learning is evident when learners demonstrate transfer of knowledge to new situations or provide multiple representations of a concept (Prain and Waldrip 2004; Bransford et al. 2000). Assessing language learners' conceptual understanding is difficult for several reasons:

- If assessment requires re-representation of an idea to a more language-intensive format, ELLs may not be able to generate the language to represent those ideas. For example, if students learn ideas through a laboratory activity and small group discussion, they may have difficulty when assessed in a paper-pencil format.)
- Language learners may repeat terms and express concepts in the new language, but still have fundamental misconceptions intact and unchanged in their prior language.
- Assessments may assess students' language proficiency rather than their content understanding.

In all subject areas, including science, teachers who work with language learners must be aware of how ELLs learn and of the challenges in teaching and assessing students' science learning. When making assessment decisions, teachers must be aware of the limitations of assessment types and adjust their strategies to determine the extent of students' knowledge.

Using Standards to Assess Learning

The challenges may seem daunting, but hope exists. The good news for science teachers is that standards documents such as *Benchmarks for Science Literacy* (AAAS 1993) and *National Science Education Standards* (NRC 1996) have been conceived as big ideas and discourage isolated facts and excessive vocabulary. Reform efforts in science education emphasize that students should express conceptual understanding rather than recall of facts and that they should express their understanding in a variety of ways. As part of accountability requirements, many states are providing options for districts to use alternative assessment formats to document what students know. In addition, teachers typically are free to design their own formats for unit exams, homework, and other assessments.

Science standards serve as an excellent starting point for designing units and assessments. The standards can be used to help organize a unit around big ideas and concepts the students are to learn. Then the textbook and other resources can serve as supporting materials. This is in sharp contrast to a curriculum organized around disjointed topics, such as assembling a series of science kits without meaningfully connecting units together or using the table of contents of a textbook to determine what to teach. Such approaches tend to overemphasize breadth of content rather than the deep conceptual understanding of fewer big ideas envisioned by the standards. Assessments, therefore, should align with the big idea—the standard—students are expected to learn and should be congruent with the instruction they have experienced.

As an example of how standards can be used to structure units and assessments, consider the topic of life cycles that most elementary science curricula include. Typically, teachers plan a butterfly unit, and young students learn *egg, larva, pupa, adult* and use these words to label drawings of eggs, caterpillars, chrysalides, and butterflies while some live insects are observed in the classroom. Unfortunately, the take-home message these children usually receive is the vocabulary words without an understanding

of life cycles, either as they apply to butterflies or to all living things. When one parent approached her first-grade daughter after a butterfly unit and asked her about what she was learning, this is how the conversation went:

Mother: What did you learn in science?
Child: Butterflies.
Mother: Oh. What did you learn about butterflies?
Child: Metamorphosis.
Mother: What's metamorphosis?
Child: Umm … Larva. Pupa … [long pause]
Mother: What do you mean?
Child: I don't know. It's like a caterpillar.

When questioned about her understanding, the child went immediately to the vocabulary of the unit rather than the concrete experiences she had had with the butterflies. When asked what the vocabulary words meant, she could not articulate her understanding. Unfortunately, the big idea that all living things have a life cycle consisting of birth, growth, reproduction, and death was lost in the translation from standards document to classroom instruction to student understanding.

Science standards, when phrased as big ideas (one-to-two-sentence statements of the central concept) should be the conceptual focus of the unit. When instruction is focused on helping students understand big ideas, assessment should then be aligned to assess students' understanding of those ideas. Students quickly learn that what is valued is what is assessed. They will attend to those aspects of instruction assessed by the teacher. If a teacher emphasizes conceptual understanding during instruction but assesses only recall of facts and terms, students quickly learn to focus their attention on learning facts and terms.

Guiding Principles for Assessing Language Learners in Science

The following principles provide a framework for our approach to assessing ELL students. They are guidelines for creating a coherent plan to document students' understanding and skills in science.

1. Know the students. ELL students likely have different cultural backgrounds that influence the way they understand and express concepts. Greenfield (1997) interviewed children and found that changing what appeared to be minor wording in a question elicited dramatically different responses—for example, changing "Why *do you think* it is the same?" to "Why *is it* the same?" Because ELL students are not a homogeneous group, getting to know them is key to addressing their needs. This entails not only interacting with students but also understanding their cultural group's ways of knowing, procedures, expectations, and assumptions. "Student perceptions of what science items are about, what they believe they are expected to do, and what problem solution strategies they use to solve them" are important aspects to consider when designing assessments for ELLs (Solano-Flores and Nelson-Barber 2001, p. 555).

2. Use multiple assessment types. Determine how the learner is putting together ideas when you assess an ELL's conceptual understanding. Because of the difficulties posed by learning a new language and having prior knowledge in a different language, look at more than one way to access a language learner's thinking. Consider coupling a written assessment with an assessment that uses relational diagrams—such as concept maps or Venn diagrams, drawings, or model construction.

3. Know the strengths and limitations of each assessment type. Each type of assessment has advantages and disadvantages when used with language learners. Although no single assessment is perfect, you can offset the weaknesses of one assessment type by coupling it with another assessment type.

4. Consider assessment an ongoing process rather than summative only. Assessment includes your observations of students while they are doing laboratory work or speaking with peers. It includes homework, group work, questions posed in class, laboratory reports, and informal conversations with students about their thinking. Documenting your ongoing observations and assignments provides valuable information about students' understanding and can be far more informative than an end-of-unit test.

5. Give language learners options when you can. Learning a new language is a cognitively and emotionally taxing endeavor. Language learners are often adjusting to a new environment at the same time, and their new school culture may be at odds with anything familiar. Testing students in a language they do not understand about new and challenging science concepts can frighten them. You can provide language learners with options for expressing their understanding to reduce their anxiety, decrease language demands, and thus improve their performance. The options include providing students more time to complete an assignment, allowing students to use their primary language when they cannot yet express ideas in the new language (provided that a translator is available), and allowing students to select the format in which their understanding is expressed—such as a model, drawing, or oral explanation.

6. Match the level of support provided in an assessment with the language proficiency of the student. Students w ho are completely new to the language of the classroom will require more support than students who are at a higher level of language proficiency. Students new to a language might use drawings or models, for example, and those at higher levels of proficiency might annotate their drawings.

7. Document modifications made for language learners. Be prepared to describe what modifications were made for language learners. Most science lessons can accommodate modifications that enable language learners to learn the same concepts as their peers yet let them express those ideas in different ways or at different times.

8. Match assessment with instruction as much as possible. Although learning is characterized by transfer, transfer from a hands-on format to a written format may be particularly difficult for language learners. Consider having language learners (and all students) transfer their new understanding to a new hands-on challenge, a visual format, or another assessment type that closely matches how students learned the new information. The key is to engage students in solving new challenges requiring the use of the new information rather than repeating new information in a different format.

9. Rethink the role of summative assessment. The end-of-unit test continues to predominate in science classes and plays a strong role in grading. Few opportunities are provided for students to express their understanding of the concept weeks or even months later. If we want students to deeply understand content, what is important is that they understand it, not when. Consider providing students more time and/or opportunities to express understanding. A single school year makes a dramatic difference in the language proficiency of a language learner. What a language learner was unable to express in October might be easily conveyed in March or April.

Applying Assessment Principles in the Science Classroom

The principles set out guide the creation of assessment plans and assessment tools that help us understand the academic performances of English learners. In the science classroom, ELLs face two learning tasks: they need to understand the science content in the lesson and the language associated with that content. You as a teacher must assess students' understanding of both the content and the language of science (Hurley and Blake 2001).

The language of science includes both vocabulary specific to a concept or lesson, such as *cell* or *mitosis*, and functions related to how that vocabulary is used. Examples of common language functions used in science are discussed in Chapter 4. As students engage in learning activities in the science classroom, these functions, such as *analyze, classify,* and *describe,* are realized across language modalities, that is, in oral language production (speaking and listening) as well as through literacy tasks (reading and writing).

Table 1 illustrates the intersection of a variety of science topics and language functions that you can integrate for effective instructional and assessment practice. The same language functions can apply across a range of topics. For example, students can record observations during a lesson on life cycles as well as during a lesson on energy. The table crosses language modality as well, because students can draw on oral and/or literacy skills to carry out language functions associated with specific science content. They can describe an organism, for example, either by answering a teacher's question orally or by writing their response in a lab report.

Our approach underscores the active role of teachers in developing and implementing assessment plans in their classrooms to collect useful information about all aspects of their students' science learning (Katz 2000). These data can help you make informed decisions about student progress in meeting science standards and the effectiveness of specific lesson content or instructional approaches. Carefully planned classroom assessment can also involve students in the assessment process so they monitor their own progress in learning and using science concepts.

Table 1: The intersection of language functions and science topics

	Science Topics					
	Change	Cycles	Systems	Weather	Organisms	Energy
Language Function						
Match symbols or illustrations to words, phrases, or sentences						
Classify or categorize objects or illustrations with text						
Hypothesize or ask questions						
Define or describe scientific events						
Record observations						
Construct graphs or charts showing data						
Analyze and interpret data						
Answer questions or report findings						
Draw conclusions						

Planning for Assessment

Good assessment is part of good teaching and emerges from careful and systematic planning. Figure 1 below sets out a four-step model for assessing student learning. Let's look at what these steps entail, especially for ELLs.

Step 1: Identify the learning standards that you want to assess. This first step in planning entails a careful analysis of the student learning to be documented through assessment. Standards are general goals for achievement and often describe broad content areas. For example, a life science National Science Education Standard for grades 5–8 focuses on structures and functions in living systems (NRC 1996). Clarify such a Standard using more specific language so that students and parents—and you yourself—understand what students need to learn. Some educators use the term *objectives* to refer to more specific targets for student achievement. Be-

cause learning involves multiple objectives and, in the case of English learners, language as well as science standards, use planning activities that will ensure comprehensive and rich information is collected. Here are some guiding questions to help structure this step:

- What science standards have been chosen?
- What specific learning objectives can be identified as stepping stones to achieve those standards?
- What activities will students engage in to demonstrate progress in achieving these objectives?
- What language requirements are embedded within the science activity?
- What cultural issues are important to consider related to this standard?
- Are the assessments aligned with the targeted learning objectives?

Figure 1: Steps for planning assessment

Step 1:
Identify learning standards

Step 2:
Design assessments to collect and record student performances

Step 3:
Set scoring criteria and methods to interpret data

Step 4:
Use information and communicate results

- Will assessments collect process or product data?

An illustration of this planning process for a middle school classroom is in Table 2.

When planning for assessment, ensure that student understanding is documented multiple times. This will show patterns of development throughout the semester and aid in instructional planning so that adequate time is devoted to each unit. In line with other principles of assessment, sample a range of objectives and collect both ongoing and summative information about students' developing science learning. Ongoing assessment makes difficulties—in student understanding, language, or in the assessment itself—become apparent earlier than is typically the case with end-of-unit tests.

Step 2: Design tools to collect and record student performances. You can use a variety of assessment approaches to collect information about students' developing competence. This is particularly effective in classrooms with English learners who may not have developed sufficient English oral and literacy skills to communicate their knowledge and understanding of science content at the same level as their English-only fellow students. Thus, as illustrated in the guiding principles of ELL assessment, students must engage in a variety of assessment activities if you are to feel confident that you are sampling students' competence in science with the least amount of interference from English language proficiency. Some of the more traditional ways you can assess student understanding are

- homework,
- lab reports,

- learning logs/science notebooks,
- reflective journals, and
- quizzes and tests.

You can structure other formats, less dependent on formal writing skills, in which students can demonstrate their grasp of science concepts. These include activities such as

- creating models,
- designing relational diagrams such as concept maps or Venn diagrams,
- completing sentences with the support of a word bank,
- engaging in role-play activities, and
- producing posters or labeled drawings.

As part of ongoing assessment, you can also use the information collected daily in the classroom as you observe students working with others, asking questions, completing classroom tasks, and engaging with natural phenomena. Use observations to calibrate instruction to the ongoing assessment through such tools as

- checklists,
- rubrics,
- rating scales, and
- anecdotal records.

How often targeted objectives are assessed, how the information will be recorded, and who will do the assessing can be plotted out over the semester or period of instruction. Table 3 illustrates a tool for planning how to collect the assessment data described as part of the unit in Table 2 on structures and functions in living systems.

Although some teachers are used to making all of the instructional and assessment decisions, it's important to remember that, as indicated in Table 2, students can and should take an active role in assessment. For example, they can use checklists to document other

Table 2: Components for planning assessment

Science Standard: Structures and functions in living systems			
Learning objectives:	Instructional activity:	Language/Culture considerations:	Assessment:
All living systems are organized in a way that has structure and function. Structure and function can be examined at multiple levels, including ecosystems, organisms, organ systems, tissues, organs, and cells.	Activity requiring comparison of a variety of living systems, discussion of commonalities found in each. Introduce differences between structure and function.	Identify language functions necessary for the science activities: Comparing Contrasting Summarizing Describing Hypothesizing Explaining	Design an organism or ecosystem. Identify structures and their functions.
All organisms are composed of cells. Most organisms are single cells. Others are multicellular. Cells carry on the many functions needed to sustain life. Cells grow and divide This requires that they take in nutrients that they use to provide energy for the work cells do and to make materials the cells need.	Drawing of ideas about smallest living things. Microscope lab with plant and animal slides. Video on biodiversity—show clips related to single-celled organisms. Generate ideas of what any living thing needs to do to survive. Cell lab and video—compare high quality cell images and textbook descriptions of what cell parts do. Groups assigned to a cell part to research and present to class.	What prior knowledge do ELL students have about structure and function? How do their cultures view living organisms (e.g., holistic vs. atomistic perspectives) and how might this affect their learning and how they express their ideas? Review language text structures: Writing lab reports Making presentations Introduce/review vocabulary specific to the activities.	Formative analysis of student drawings—their ideas about what the smallest living things look like. Watch for differing cultural perspectives. Build a labeled cell model with description of structure and function of cells. Or, group presentations of cell part, with each part describing how it interacts with other parts.
Specialized cells perform specialized functions in multicellular organisms.	Students draw what they think human cells look like from skin, brain. Microscope lab with variety of human cells from different organs. Discussion of why the body might have different cell types.	Demonstrate graphic organizers: Labels for diagrams of cells Semantic maps for comparing living systems	Observational record of student ideas expressed during discussion and student cell drawings.

students' performances or to assess their own work.

When you carefully prepare assessment plans, you can check whether you are indeed varying your approach to assessment by using multiple methods, engaging students in self- and peer-assessment, using both oral and written modalities to check understanding, and collecting both formative and summative data.

Step 3: Set scoring criteria and methods to interpret data. Because of the typical language load of traditional assessments such as paper-and-pencil tests, alternative assessments are often used to determine the level of achievement of English learners in the science classroom. Although looking through a portfolio of student work can provide insight into student achievement, you

also need to understand what an individual student performance means and how it can be weighed or measured against explicit criteria. Clear and explicit criteria provide students with information on what counts in assessment and ensures that you consistently assess student work across students and across performances.

Two common ways of understanding student performances are by comparing individual performances against a norm or by comparing them against an expected outcome based on the aggregated performances of other students on similar tasks. Many standardized tests use a norming population as the basis for establishing levels of performance against which students' individual scores are measured. The Iowa Test of Basic Skills (University of Iowa) is an

Table 3: Plan for collecting assessment information

Assessment activity	Individual, small or whole group assessment	Recording tool	Person collecting data
Design and label an organism or ecosystem	Individual	Checklist	Students working in pairs review each other's structures
Present cell part research projects	Small Group	Rubric	Teacher assesses each group using class-developed rubric
Student drawings of what they think the smallest living things look like	Individual	Anecdotal Records	Teacher assesses during activity noting similarities, prior knowledge, key differences
Discussion of what students think human cells look like	Whole Group	Anecdotal Records	Teacher assesses during activity noting how students think about cells

example of such a test. A further way to interpret student performances is to compare the performance against criteria geared to instructional objectives. Scoring rubrics are common examples of such criteria-based assessments; they reflect the expectations for student performances.

When students are doing science, you can assess their performances using rubrics and scoring guides to record their efforts. You can design your own rubrics and scoring guides to reflect targeted learning outcomes. Here are a few questions to guide the development of these rubrics:

- How well are important science outcomes reflected in the criteria?
- Are the criteria clearly written and in a usable format?
- To what extent do the criteria reflect classroom instruction?
- To what extent do the performance levels reflect appropriate expectations for student achievement?

- To what extent do the criteria provide English learners with opportunities to show their understanding and skills?

Table 4 provides an example of a rubric used to assess student performances in completing a science project. In this case, the second-grade teacher used this tool to document the quality of student work in creating a life cycle poster.

Some teachers give two types of feedback to English learners, one for science learning and the other for their use of English to express that learning. Given both feedbacks, students can develop a clear understanding of their progress both in achieving science outcomes and in developing their academic language skills. You can give this feedback in separate grades or sets of comments or, as in the case of the rubric in Table 4, incorporate both kinds of feedback in one assessment tool.

Step 4: Use information and communicate results. Although we have focused on how

Table 4: Rubric for a grade 2 life cycle poster project

Excellent	Very Good	Satisfactory	Improvements Needed
All stages of life cycle shown in proper order, and clearly indicates that reproduction begins a new cycle	All stages of life cycle present, but conveys a linear fashion of one series of stages without showing how reproduction leads to a new cycle	Some stages of life cycle missing, or present but out of order	Life cycle stages missing and out of order; does not convey an understanding of a life cycle
Illustrates examples of 2 or more organisms, accurately labeling life cycle phases in each	Illustrates one organism and accurately labels its life cycle stages	Life cycle stages are inaccurately labeled for one organism	Does not show life cycle stages for an organism

to construct meaningful and appropriate assessments, remember that assessment is not the end point. Assessment data should be used and, in most cases, reported to others. Some assessment results, for example, will inform students and other stakeholders of individual student achievement. You may also use assessment data to review the effectiveness of instructional strategies or particular science materials. When assessment data are understandable and presented clearly, reporting of results is enhanced. That may entail reporting results in languages other than English when you communicate with parents of English learners. Keep in mind that parents of English learners may have limited understanding of their child's new educational system. They often appreciate an explanation of the context, your expectations, how to interpret assessment results, and how they can help their child. Most important, for assessment information to be useful, it must be reported in a timely fashion.

Using Assessment to Improve Learning

In recent years, assessment has come to be associated with standards and gathering data for large-scale accountability purposes. Although such data help schools and districts gauge the performance of groups of students on a large scale, such a focus should not obscure the important role of the student in assessment. Clear feedback is important for a learner as he or she seeks to improve. Further, a student needs that feedback close to the time of the performance; the longer the time delay between the performance and feedback, the less likely the learner is to incorporate

the feedback into understanding (Gagne, et al. 1993). Effective assessment should serve a dual purpose: it should provide the teacher and other interested parties insight into a student's understanding, and it should promote further learning by providing feedback to the student as well as altering the teacher's practices based on the insights gained from the assessment. Standardized tests provide information to interested parties about students' achievement, but rarely provide specific, timely feedback to the students being assessed.

Providing Meaningful Feedback to Students

Students typically get feedback in two ways, either as spoken comments with accompanying intonation and body language or as written comments made on papers, tests, and other written work. Be aware of how both are used by students to gauge their level of understanding and make efforts to improve.

Spoken Feedback

Spoken feedback is the most common form of feedback. Language learners may have difficulty understanding what you are saying and may rely on nonverbal cues and intonation to help them interpret the feedback. For this reason, you must be careful that the overt and covert messages match. When you announce to the class, "I like the way Table 1 is working," the real message may be that other groups are not working well. You are drawing attention to the students at Table 1, and other students may get angry with that group or feel insulted. A language learner may interpret your comment to mean that you like Table 1. A better strategy might be to approach groups directly and quietly address specific problems or accomplishments.

Reducing, or eliminating, idioms from your language is an important way to make verbal feedback more comprehensible for language learners. Because it is difficult to realize how idiom-laden our speech can be, tape yourself. Listen for phrases that might make no sense to a language learner, and consider how to restate them to aid in comprehension. Here are a few examples:

- "Your work was *top notch*, with the first paragraph *flowing* nicely into the second."
- "It seems that the food you gave the hamster really *hit the spot*."
- "Once you have your materials, you can start *right off the bat*."

Be careful to ensure that students aren't given grossly differential verbal feedback. When a group of novice teachers was shown a videotape of a first-grade lesson without knowing any of the students, they knew immediately which child was the lowest-achieving student in the class. Children also have this information, and they learn it by noticing differential feedback the teacher gives. In the tape, the teacher elicited a variety of responses from students to an open-ended question about what they knew about being healthy. When most students contributed an idea, the teacher said, "Okay, that's interesting," and wrote the child's idea on a chart. When one particular child provided an idea that was not substantially different than the ideas already provided, the teacher switched his responding pattern and said, "Oh, Katie. That's very good." Such feedback implies that the teacher has different expectations for different children and that some children get a lot of praise for doing very little. Although we certainly want to encourage our language learners, we must ensure that our responding pattern does not convey a message of lower expectations for them.

Written Feedback

Vague statements such as "Good job!" and "Way to go!" may be quick to write on students' work, but do not provide meaningful information to students seeking to improve their performance. Effective feedback, including praise, should be specific to the task, given in private, and attribute successes or failures to factors under the control of the student, such as effort or strategies employed rather than luck or natural ability (Brophy 1981; Gagne, Yekovich, and Yekovich 1993). In this way, you give students information about their understanding and they know better how to improve.

Written feedback is important for all students, but particularly for ELLs. They can consider the information at a slower rate than spoken language and can more readily determine where they were successful and where they need to improve. Providing scores to students is certainly more efficient, but be aware that a number on a piece of paper conveys far less information to a student seeking to improve. Also consider that some cultures do not assign "points" to students' work, thus rendering such numbers meaningless to those students. When you provide scores, make sure students have access to a scoring guide to know what the score means and why they were given a particular score.

Finally, written feedback can be used to further teach students (Cox-Petersen and Olson 2002). Posing questions is an important way to help students make important connections and advance their thinking. You can write thought-provoking questions such as "Why might this be?" and "How is this linked to ___?" in the margins of students' work.

Promoting Metacognition

Self-assessment can be a valuable tool for students, particularly language learners. Metacognition, or helping students think about their own thinking and progress, can trigger more thinking, particularly about ideas that are embedded in students' primary language. For example, having students write or draw their thinking in their most familiar language and then transferring it to the new language can be a valuable way to help them make important connections and become aware of misconceptions or important differences between their familiar cultures and their new culture. At the end of a unit, students can compare earlier writings and drawings to their current thinking and explain how their thinking has changed.

An additional aspect of metacognition involves goal setting. Learning a new language can make a student feel isolated, particularly if he or she lacks classmates who speak the same primary language. Working with students to set achievable, yet challenging goals helps ELLs more closely monitor their progress in both English language proficiency and science understanding over time. This can decrease the anxiety students experience from uncertainty about where the instructional sequence is going and what is expected of them. Students need access to criteria in both English and science, a schedule for the class, and time to reflect on what they would like to accomplish, with targeted dates for the accomplishments. You then need to work with students to determine if their goals are appropriate, to monitor progress, and to celebrate achievements.

Summary

Despite the challenges of assessing English learners in science, many strategies help diagnose and promote students' understanding of science concepts and English proficiency. Knowing students and their cultural norms is important to understanding why certain questions or concepts might be particularly problematic. You can use this knowledge to better structure assessment tasks as well as instruction. Using multiple assessments, particularly those with decreased language demands, is important for determining what students understand. Plan such assessments deliberately to ensure that they align with important standards and objectives and that ELLs have the opportunity to demonstrate their understanding in a variety of ways. As students increase in their proficiency, the language demands of assessments can be increased.

Consider how best to provide assessment feedback to ELLs so that they will work to improve. Spoken and written feedback are two ways that students gain information about their achievement from the teacher. Make these messages as comprehensible and specific as possible so that ELLs know where to focus their learning efforts. Finally, involve ELLs in metacognitive activities as an additional way to help them understand their new culture and learn both English and science.

References

American Association for the Advancement of Science. 1993. *Benchmarks for science literacy.* New

York: Oxford University Press.

Bransford, J. D., A. Brown, and R. Cocking. 2000. *How people learn: Brain, mind, experience, and school.* Washington, DC: National Academy Press.

Brophy, J. 1981. On praising effectively. *The Elementary School Journal* 81.

Cox-Petersen, A., and J. Olson. 2002. Assessing student learning. In *Learning science and the science of learning*, ed. R. Bybee, 105–118. Arlington, VA: NSTA Press.

Gagne, E., C. Yekovich, and F. Yekovich. 1993. *The cognitive psychology of school learning.* New York: HarperCollins.

Greenfield, P. M. 1997. Culture as process: Empirical methods for cultural psychology. *In Handbook of cross-cultural psychology, Vol. I*, 2nd ed., J.W Berry, Y.H. Poortinga, and J. Pandey, eds., 301–346. Needham Heights, MA: Allyn and Bacon.

Hurley, S., and S. Blake. 2001. Assessment in the content areas for students acquiring English. In *Literacy assessment of second language learners*, eds. S. R. Hurley and J. V. Tinajero, 84–103. Needham Heights, MA: Allyn and Bacon.

Katz, A. 2000. Changing paradigms for assessment. In *Implementing the ESL standards for pre-K–12 students through teacher education*, ed. M. A. Snow, 137–166. Alexandria, VA: TESOL.

National Research Council. 1996. *National Science Education Standards.* Washington DC: National Academy Press.

Prain, V., and B. Waldrip. 2004. Using multi-modal representations of concepts in learning science. Paper presented at the annual meeting of the National Association for Research in Science Teaching. Vancouver, BC.

Solano-Flores, G., and S. Nelson-Barber. 2001. On the cultural validity of science assessments. *Journal of Research in Science Teaching* 38: 553–573.

University of Iowa. 1935, regularly updated. Iowa Test of Basic Skills. Available online at *www.education.uiowa.edu/itp/itbs*.

Vygotsky, L. S. 1978. *Mind in society: The development of higher psychological processes.* Cambridge, MA: Harvard University Press.

Further Reading

Atkin, J. M., and J. E. Coffey, eds. 2003. *Everyday assessment in the science classroom.* Arlington, VA: NSTA Press. The chapters in this book are devoted to different timely issues related to assessment. The role of questions in assessing and promoting thinking is thoroughly addressed, and a chapter on examining student work is very helpful for science teachers, particularly those who work with ELLs.

Enger, S. K., and R. E. Yager. 2001. *Assessing student understanding in science.* Thousand Oaks, CA: Corwin Press. This practical guide provides many concrete examples of assessment types and scoring rubrics across multiple domains of science knowledge: concepts, processes, application, attitude, creativity, and the nature of science. Although many examples can be applied to multiple grade levels, additional chapters provide grade-specific examples—K–4, 5–8, and 9–12. A glossary provides helpful definitions coupled with research references related to assessment.

Hein, G. E., and S. Price. 1994. *Active assessment for active science: A guide for elementary school teachers.* Portsmouth, NH: Heinemann. This excellent resource on science assessment for the elementary grades guides a teacher through the entire process from selecting an assessment type to managing assessment, interpreting children's work, and scoring. It provides sample assessments with many examples of student work. Sections on interpreting and scoring students' work, including drawings and conversations, are very useful.

Irujo, S., ed. 2000. *Integrating the ESL standards into classroom practice: Grades 6–8.* Alexandria, VA:

TESOL. This is one of four volumes in a series intended to guide teachers of English language learners in using standards to inform their teaching. Each unit in the volume focuses on a particular content area, identifying both content and ESL standards and aligning them with lessons and procedures designed to support student learning. It provides a wealth of sample instructional activities and assessments as well as an excellent set of references. Other grade level spans include preK–2, 3–5, and 9–12.

Science Beyond Classroom Walls: Fairs, Family Nights, Museums, and the Internet

John R. Cannon, Judith Sweeney Lederman, Monica Colucci, and Miosotys Smith

The school bus filled with buzzing in English and Spanish as it approached the downtown area. This was the day that the students had been waiting for —a field trip to the city planetarium. Faces were glued to the windows as the bus parked in front of the massive building. Ms. Thompson, the Earth science teacher, said, "Este es el planetario," followed by "This is the planetarium," and then told the class about what they were going to see in the planetarium and what behavior she expected of them. As she was speaking, Ms. Thompson first made her comments in Spanish, then translated to English.

Excitement was high as the class exited the bus and began climbing the long flight of stairs leading into the building. "Este edificio es enorme," (this building is huge) said one student. Once inside, the class was amazed to find a huge plumb bob at the end of a long cable attached to the ceiling, swinging back and forth, knocking down small pegs placed in a circle. The class was buzzing again, this time with many questions. "¿Qué esta cosa divertida?" said one child. "Well, Valentina, this funny thing, or 'cosa divertida,' as you call it, is really a pendulum," Ms. Thompson answered. "Puede ser utilizado para marcar el tiempo" she went on. "It can be used to tell time." As the tour continued, Ms. Thompson did her best to hear each question and to answer it in both English and Spanish.

Window
Into the
Classroom

Ms. Thompson, a first-year bilingual science teacher, was not surprised that her tenth-grade Earth science students had never visited the city planetarium. Most of her students were newly arrived from Central and South American countries; on average, she welcomed one a new student per week to her classroom. Earlier in the semester she had learned that few of her students had ever seen the lake that was just two blocks from the high school. Most had never been in a museum of any type.

Ms. Thompson had taken a course in

informal education as part of her preservice education program. The course included a practicum at an informal site, and she chose to work with the education department at the planetarium. She observed many student field trips and witnessed firsthand the benefits of connecting classroom instruction with the resources of the planetarium. Now that she had a class of her own, she was anxious to make these resources available to her students. She also wanted to give them a broader glimpse of their new home by taking them into the city. Her students were studying Earth–Sun systems and the solar system at the time she arranged for their field trip. Because she was already familiar with the sundial and solar system exhibits at the planetarium, it was easy for Ms. Thompson to design a field trip that reinforced the concepts and vocabulary that were part of her formal classroom instruction. She thought her students would be fascinated by the animated planetarium show and would find the exhibits engaging.

She was not prepared, however, for her students' reactions when their bus pulled up to the planetarium and they saw the giant sundial in front of the building. They were absolutely delighted that they not only knew what it was but that they also were able to use it to determine the time. They expressed a sense of intellectual pride. Inside, it wasn't the flashy planetarium show that interested them but rather the armillary sphere in the lobby that caught their attention. Once again this was something they had studied in class. What they had studied about was what they enjoyed the most.

The next day Ms. Thompson had her students write about their experience in their science journals. They all said they wanted to go back. But most of them wanted to go back with their families to share this experience with them. One student's journal entry translated into this:

I liked the museum a lot. There were many things there I did not know about. But there were also things I did know about already like the sundial at the entrance to the museum and the solar system. I was surprised I knew so much! I also liked the view of the city and the beautiful lake. I learned about how people thought about the Earth a long time ago. They believed that the Earth was the center of the universe and that the other planets were going around it. They realized that the Sun was at the center and the planets, including the Earth, were going around it. I would like to go back to the planetarium again with fewer people so I can have more time to look at things. I would like to bring my family next time. My sister and brothers would like this place and I could show my mother and father what I have learned.

Several positive outcomes, both intended and unexpected, resulted from the field trip. The students found the planetarium to be a novel, exciting learning environment. It augmented the science content they were learning with media and exhibits not found in their formal classrooms. But, most important, it provided an opportunity for the students to realize they already knew some science and to identify themselves as successful learners. That made them proud of themselves and eager to share their experience with their families.

Informal Science Learning Experiences and Resources

A field trip is only one example of how valuable connections between formal and informal education settings are for the teaching and learning of science. Informal science education involves all nonclassroom-based learning experiences that make use of community resources and expertise for K–12 students and teachers. Informal education can take place in large national and regional institutions such as museums, aquaria, botanic gardens, zoos, and science centers as well as small neighborhood-based community centers, libraries, and parks. The National Science Teacher Association's position statement on informal science education (1999) "recognizes and encourages the development of sustained links between the informal institutions and schools." The National Science Education Standards (NRC 1996, p. 45) acknowledge that "the classroom is a limited environment" and that "school science programs must extend beyond the walls of the school to the resources of the community." It recommends that district and school leaders "provide opportunities for students to investigate the world outside the classroom" (p. 46) and suggests creating collaborations that link "the best sources of expertise" (p. 51) with "the experiences and current needs of teachers" (p. 50).

Recognizing these needs, informal institutions have designed a variety of educational opportunities that support and augment formal K–12 science instruction. These include inclusive, learner-centered, and content-rich experiences and resources for students, as well as high-quality professional development programs for teachers. Many informal sites waive admission charges for teachers to encourage them to become familiar with the site before bringing their students on trips, and some offer professional development programs to help teachers design field trips that focus on single exhibits or groups of related exhibits that best connect with the teachers' curricula.

The goal is to have teachers use the informal site as an extension of their classrooms. Teachers are often invited to special events to preview new exhibits and receive copies of related educational materials. Many sites offer services, programs, and curricula materials in several languages to accommodate the cultural diversity of their local communities. The Exploratorium in San Francisco, the Brookfield Zoo in Chicago, and the Franklin Institute in Philadelphia will arrange for translators for visiting school groups. The Exploratorium's Online Exhibit website at *www.exploratorium.edu/exhibits* can be accessed in Spanish, French, and Italian. The Museum of Natural History and planetarium in Providence, Rhode Island, will provide written translations of its planetarium shows to school groups upon request. The Fort Worth Museum of Science and History has several bilingual exhibits that present signage in both Spanish and English. Like many other informal organizations, it strives to align its K–12 education programs with state and national standards.

In addition to on-site programs, informal organizations offer outreach programs that take place directly in schools, libraries, and community centers. The Brookfield Zoo presents its Zoo Adventure Passport program in both Polish and Spanish at local schools and libraries. Related science trade books in English, Spanish, or Polish are available for attending children. The Brooklyn Aquarium in New York City often has translators at its school-based outreach programs and has conducted lessons in Russian, Chinese, Hebrew, and Spanish. Chicago's Field Museum of Natural History and the Boston Museum of Science have outstanding collections of lending boxes that address an array of science topics. Each box contains artifacts and inquiry-based curricula. Some include classroom sets of hands-on trade materials and grade-appropriate trade books.

Long-term initiatives are other examples of collaborations being developed between informal sites and local community centers. The Chicago Botanic Gardens and Gads Hill Center, a nonprofit family resource center serving the predominately Mexican American Pilsen community in Chicago, have collaborated on Primero la Ciencia: An Urban Environmental Summer Camp. This program is designed to nurture interest in science and scientific inquiry through experimental learning. It targets neighborhood middle level students and consists of summer

and school-year programs. All materials and instructions are offered in both English and Spanish. Lessons are conducted at the community center, neighborhood parks and lake, and the Chicago Botanic Gardens. The goals of this program are shown in Table 1.

The Museum School within the Fort Worth Museum of Science and History has a similar community initiative in partnership with a local college. Funded by a grant from the National Institutes of Health, it has developed programs for three age groups of underserved Hispanic students. The school runs summer camps that promote the learning of inquiry-based science. Just as with the botanic garden project, one of its goals is to introduce the students to science as a possible career. The group has a participating teacher who works throughout the school year with the students to support their interests and learning in their formal science classrooms.

In some ways, the problems confronting teachers of English language learners (ELLs) are no different from those in any diverse classroom, including classrooms with students with special needs. Often, barriers prevent the teacher from making instruction as

Table 1: Goals of Chicago Botanic Gardens summer camp

- Increase students' knowledge of plants and plant ecosystems;
- Introduce students to inquiry and problem solving techniques that have applications in science and across disciplines;
- Increase students' understanding of the scientific process;
- Increase students' confidence in doing science on their own and in collaboration with others;
- Introduce students to careers in the natural sciences; and
- Give students a better understanding of the role of botanic gardens.

relevant as it can be to all students. Barriers can also prevent the teacher from accurately assessing what students know about a particular topic. Quite simply, classroom conversation is not necessarily the best way to connect with students or to find out what they know. In addition to obvious barriers of language, numerous interpersonal and social issues compromise instruction. Informal education venues bring students back into their familiar communities and make the connection between the school curriculum and students' lives more concrete. Students whose cultures and native language differ from those of their teacher and classmates can be provided with a venue for learning and emotional development that transcends such differences. Informal education venues help bring out common experiences and knowledge that already exist beneath the surface in students brought together in today's diverse classrooms. Breaking down the barriers created by language and culture can only facilitate the process of teaching and learning.

Overall, the goal of science instruction, and education in general, is to maximize the learning of all students. This goal is often compromised by cultural, societal, and language differences. The value of informal education venues is that they can help a teacher overcome such barriers and still celebrate the diversity among students that enhances the educational process. Another way to extend the walls of the classroom is to look at schoolwide programs that focus on science. Two of the more popular programs that occur in many schools are the science fair and its nonjudged counterpart, science festivals, and family science night program.

Science Fairs and Family Science

Toward the end of the school day, the children at Crestview Elementary were busily taking care of the final details before that evening's Science Fair and Family Science Night. Crestview Elementary has been holding family science nights in conjunction with the school's science fair since the mid-1980s. The staff found that, by combining the two events, more students and parents became involved in each. The multipurpose room was arranged in two sections: one section housed the Science Fair projects on display and the other was filled with activities that families would soon be doing during the Family Science Night portion of the evening. Tables were filled with balloons, bamboo skewers, and many other materials. As the children left, one could see and hear their excitement because of the evening of science that was just a few hours away.

Involving parents of ELL students in the education of their children is particularly important because it helps strengthen interpersonal communication skills in the native language. These skills include asking questions, making observations, making inferences, and thinking critically and creatively in the nonthreatening environment of the home. Furthermore, parental involvement helps validate the students' experiences, values, and cultural backgrounds. This validation effectively and constructively affects students' confidence and ability to communicate in a new language. Parental support helps ELL students develop an appreciation for learning and consequently acquire cognitive academic language proficiency.

Science Fairs

The activities discussed here took place in an urban school in South Florida that services students in preK–8. The student body of approximately 1,300 was 89% Hispanic (the U.S. Census Bureau reports that, in 1997, approximately one-half of the country's foreign-born population were from Latin American countries), 1% Black, 9% percent White Non-Hispanic, and 1% percent Multiracial. Of these, approximately 20% were ELL students. Since Spanish was the predominant language of the students and their families at this school, a bilingual program in the content area focused on strengthening students' native language literacy and academic skills, particularly in the lower grades, as they gradually learned a new language.

Many schools in the United States conduct an annual schoolwide science fair. The science fair project typically involves scientific inquiry: students are required to identify a problem, hypothesize the effect of a treatment upon a variable resulting from the problem, and develop a plan to carry out an investigation to support or refute their prediction. Furthermore, they must interpret and draw conclusions about their results. In the school described, this event is particularly valuable in helping ELL students advance their language acquisition and their academic skills through teacher support and parental and family involvement. A Science Fair Program Committee is established at the school site to discuss and implement the approach and strategies that will best help these students and their families to effectively participate. The program takes one grading period of ap-

proximately nine weeks toward the end of the school year. A Science Fair Exhibit culminates the event. A committee judges each project using a rubric that is also provided to students. All students are recognized for their efforts, and a first, second, and third place is awarded to students in each grade level. The intensity of such a project, which is basically developed at home, requires strong parental involvement to support the learning that takes place in the classroom.

Parents of ELL students may find helping their child with the process and completion of a science fair project quite cumbersome, because their command of the English language may be even more limited than that of their own children. Support from the school is instrumental. The science fair program begins with a bilingual Science Fair Parent Night where parents are walked through a parent- and student-friendly guide, prepared in English and Spanish, to help them assist their children in developing a science project. This guide contains step-by-step explanations and examples of how to develop a science project using a scientific process. It includes forms for each component of the project and a calendar that specifies the expectations and preliminary due dates of each component of the project to provide guidance throughout the process. A rubric, also included, presents a way for the students and parents to pre-evaluate each part of the science project as they work together through its development.

During this bilingual Science Fair Parent Night, priority is placed on validating the non-English-speaking parents' cultural backgrounds and the value they bring to their children's education. It is emphasized that using their native language in the discussions at home will help their children think through the process and develop their projects. Parents are provided with ample opportunities to ask questions and express concerns, all of which are addressed in their native language, in this case Spanish. During the family night, a demonstration of how to do a simple hands-on experiment reinforces what is presented in the Science Fair Guide. Obviously, all information and documents related to the Science Fair—such as flyers, pamphlets, and guides for parents and students—are translated from English to Spanish to eliminate intimidation resulting from language barriers. The students present the drafts of each assignment to the teacher for revision and final approval. This process cultivates and enhances the students' vocabulary and composition skills in their developing new language.

Spanish, the primary language of most students and families in this school community, offers an additional opportunity to link science terminology to the native language. Many scientific words are formed from Latin and Greek words, and Spanish is directly influenced by these languages. Words like *observation, control, experiment, results, evidence,* and many others within the content and practices of science are conceptually, visually, and phonemically familiar to Spanish native speakers.

The science fair program ends with a science family night. All parents and students are invited to the school to view the exhibit of completed and judged science fair projects. Visiting the Science Fair allows parents and their children to question, discuss, and analyze the projects of others, further rein-

forcing language skills and development. At the fair, ELL students can share what they have learned and experience a sense of pride and accomplishment as they see their projects displayed. This positive experience promotes enthusiasm for learning.

Providing links between the school and home is most valuable to the ELL who encounters the difficult task of mastering content through instruction while learning a new language. The schoolwide science fair stimulates the ability of the human being to develop language. Science deals with the concrete and the factual. It is from the same perspective that language develops and evolves, concepts awaken, and words are born to satisfy our instinctive need to construct knowledge of the world and communicate our thoughts and our understanding.

The Family Science Program

Family Science (Foundation for Family Science 1999) is an informal science education program that gives parents and children an opportunity to work and learn together while doing hands-on science activities. The program began in 1988, modeled after the popular Family Math program designed by the EQUALS program at the Lawrence Hall of Science, University of California, Berkeley. The Family Science program, headed by David Hiel and Associates in the Foundation for Family Science, has three goals:

- making science more accessible to families,
- demonstrating the relationship between science education and future career choices, and
- getting parents more involved in their children's science education.

Although the Family Science program

had been promoted as a resource for teachers and parents, it now guides family science nights at which parents and children return to school during the evening for fun in science within the family unit. Family science nights, together with science fairs, give children of all cultural backgrounds the opportunity to experience science. For more information about family science nights, please visit the Sandia National Labs at *www.sandia.gov/ciim/ASK/html/elementary/family-night.htm* and *www.sandia.gov/ciim/FSN.*

Technology: Resources for Teachers and Students

Today is the day students in Ms. Davis's class have been looking forward to all week—the day they visit the school's computer lab. The class comprises fifth-grade students from all walks of life and diversities. Although English is the primary language spoken in class, three other languages also can be heard from time to time. This multilingual classroom composition is typical for Ms. Davis's school.

Window *Into the* Classroom

Computer time in the lab today will be spent on a WebQuest, an inquiry-oriented activity in which most or all of the information used by learners is drawn from the web. It is designed to use learners' time well, to focus on using information rather than looking for it, and to support learners' thinking at the levels of analysis, synthesis, and evaluation.

Ms. Davis's class will embark upon Dr. Green's Rainforest Mystery, a WebQuest about animals living in the rainforest (San Diego State University WebQuest available online at http://webquest.sdsu.edu). This electronic expe-

dition involves using the late doctor's field notes and a photograph to identify an animal never seen before. This is not the first WebQuest for the students: Ms. Davis uses WebQuests for approximately half of her science instruction, while the remaining time is spent on hands-on, activity-based lessons.

The students are eager to begin. Learning teams, made up of two students with one perhaps being an English language learner, begin reading aloud from the computer monitor, using "12-inch voices" so as to not to disturb the other teams. ELLs follow along with their partners, as words are being spoken and pointed to on the screen. It says:

"You and your research team are responsible for reviewing Dr. Green's notes in order to find the name of this animal, the types of foods it eats, and which rainforest it lives in. To inform the rest of the Animal Protection Center about this animal, you must create a brochure describing all of the information you have researched. Using this information, create a 3-D model of the animal and its habitat that could later be used to design an appropriate living environment for the animal if it were to live in a zoo. Create an informative poster that supports your opinion about whether the animal needs protection."

As the students start their journey, the English-as-a-second-language teacher comes through the door and walks over to a learning team, kneels down, and asks, "What are you two doing today?" Both students explain, as best they can, the task at hand. She pulls over a chair, reviews the task with the team, and asks, "Well, then, where should we go first?" The ELL team member points to a hyperlink on the screen. The English language development teacher asks the ELL to repeat the hyperlink and helps with pronunciation. A smile comes over the face of the child, and the teacher gives each student a high five as she makes her way over to another learning team.

Technology in education continues to be at the forefront of innovation and curricular change, akin to programs that extend the walls of the classroom. Technology and the use of the internet significantly increase opportunities and resources for both the classroom teacher and his or her students. In 2004, it was estimated that more than 68% of the population in the United States uses the internet in some capacity (Internet Usage Statistics 2004).

WebQuests, as used by the English language learners in the "Window Into the Classroom," are rapidly becoming part of classroom activity in all subject areas. The Australasian Society for Computers in Learning in Tertiary Education (ASCILITE 2000) reports that

> A good WebQuest represents good pedagogy. As it was conceived by Bernie Dodge and Tom March, a WebQuest is a curriculum-based inquiry activity with well-chosen internet resources within well-defined tasks. The inquiry orientation requires high order thinking skills applied to the challenging questions presented to the students, and then solved using the best of internet resources for a particular curriculum area. The structure of a WebQuest generally fosters collaboration, and at least one outcome calls for the presentation of the students' newly acquired knowledge, sometimes in an interactive multimedia format. As students meet the challenges of WebQuests, they use and develop a variety of technology skills.

WebQuests represent the best in emerging inquiry-related innovations for the 21st century and directly reflect both the National Science Education Standards' (NSES) (NRC 1996) definition of inquiry and support the five essential NSES features that must be present for inquiry to take place while using technology. These features are as follows:

- Learners are engaged by scientifically oriented questions.
- Learners give priority to evidence, which allows them to develop and evaluate explanations that address scientifically oriented questions.
- Learners formulate explanations from evidence to address scientifically oriented questions.
- Learners evaluate their explanations in light of alternative explanations, particularly those reflecting scientific understanding.
- Learners communicate and justify their proposed explanations.

No longer are student investigators, or researchers, bound by classroom walls. Access to the internet allows children of all ages and cultural backgrounds to be engaged by and explore scientific questions, collect electronic data, use the data to develop explanations related to their investigations, and propose and communicate explanations to others. Although this form of investigation lacks physical manipulation of materials, it stimulates the minds-on aspect of science education that, at times, has been lacking in hands-on, activity-based science programs and curricula.

Schools can and do expose children to technological innovations as best as they can, thereby capitalizing on learning in both science and language. Whether with one computer lab for the school or one computer for each child, children are being taught how to better use technology in their own education.

Computer-assisted language learning (CALL) has been greatly enhanced by the advent of the internet, World Wide Web, and multimedia applications. Students now can find information, download information in a multitude of languages, and communicate with one another via e-mail, thus closing learning barriers and expanding the borders of knowledge. This technological revolution has benefits for the English language learner and the science classroom. Phinney (1996, p. 152) found that "the variety of 'documents' my students have made using the computer has engaged their interest, motivated them to use the second language (English) outside the classroom, and helped them to learn the cooperative skills they will need in the working world."

For the English language learner, "Technology can provide one of the most powerful sets of instructional tools for teaching to content benchmarks. Furthermore, technology can be used to provide ELLs with extremely effective scaffolding, which makes instruction accessible and can help to create a dynamic environment for direct language teaching" (Phoenix Talent School 2003).

Websites for Teachers and Students

The online resources in science and in language available to students and teachers are numerous. When used in conjunction with learning science and building language proficiency, the World Wide Web can be an indispensable tool.

Many of the sites that follow have large collections of hyperlinks to science content and science teaching websites.

Even though the following list has been broken into differing sites for students and teachers, they are not exclusive. Students and teachers can learn new and exciting information from any site on the list. Take some time to review them, and always be watching for those sites that are the proverbial jewels in the rough.

Web Resources for Students

Ben's Guide to US Government for Kids

http://bensguide.gpo.gov/subject.html

Don't let the title fool you. Inside, you'll find links to sites related to science, history, and yes, the U.S. government. While you're there, don't forget to take a virtual tour of our National Zoo. It is a 163-acre zoological park located within Rock Creek National Park in the heart of Washington, D.C. The zoo is home to more than 2,700 individual animals of 435 different species. The best-known residents are the giant pandas, Tian Tian and Mei Xiang, and their child.

The Why Files

http://whyfiles.news.wisc.edu

People sometimes think of science as remote, relevant only to a group of white-coated people tucked away behind laboratory doors. This site presents science for what it is: an important human enterprise performed by people. Topics and issues include science, health, environment, and technology, all from a unique perspective.

Science 4Kids

www.ars.usda.gov/is/kids

Supported by the United States Department of Agriculture, Science 4Kids has links to information on subjects ranging from animals to weird science. This is a superb collection of agriculture-related science sites.

Kids URLS

http://research.esd.ornl.gov/%7eforrest/kids-urls.html

Forrest M. Hoffman, a team member at the Oak Ridge National Laboratory in Tennessee, has amassed a huge number of sites to visit. There are links to many important websites, including NASA, the Franklin Institute Science Museum, and the Virtual Frog Dissection Kit.

Bizarre Stuff You Can Make in Your Kitchen

http://bizarrelabs.com

Brian Carusella has created a museum of classic home science experiments, mainly from the 1930s to the 1960s. Many of the really good science activities and demonstrations from the past are still as good today. Activities like Gum Wrapper Poppers and Stupid Egg Tricks are just a few that you'll find at this site.

Ask Dr. Science

www.ducksbreath.com

The Ask Dr. Science website claims to be "the home of America's foremost authoritarian on the world around us. Or at least the world around him." It goes on with "Ask Dr. Science … he knows more than you do." Although the site is a bit tongue in cheek, it gets questions answered.

The Yuckiest Site on the Internet

www.yucky.com

There has to be a website that can answer personal questions, like those about body functions. Click on the "Gross and Cool Body"

link to learn more about how people sweat, get zits, produce ear wax, and yes, even digest their food. You'll learn a lot about science.

How Stuff Works

www.howstuffworks.com

How Stuff Works helps readers eliminate the mysteries about how things work through its creator, Marshall Brain, and his ability to communicate complex ideas clearly. Find the answers to everything from "how toilets work" to "how atoms work" on this outstanding resource for teachers and students.

Bill Nye, the Science Guy

www.billnye.com

Included are features such as the Demo of the Day, Home Demos, and episode guides that accompany each of Bill Nye's television programs. The site also includes the Bill Nye classic question-and-answer segment "Did You Know That …"

Yahooligans! The Web Guide for Kids

http://yahooligans.yahoo.com

Yahoo has produced one of the best one-stop search engines for kids—Yahooligans. One would be hard-pressed to find something of interest to children missing from this site. There are links galore to everything from the core school subjects (science, math, and language arts) to the Joke of the Day.

Web Resources for Teachers
The Smithsonian Institution

www.si.edu

The site has many sections that will add science, technology, and historical content to anyone's background, novice to expert.

National Geographic Society

www.nationalgeographic.com

Launched in 1996, the National Geographic Society website includes the sections National Geographic World, Explorer, and story highlights from the *National Geographic* magazine. Other sections provide many teaching resources and the National Geographic Map Machine, on which you can click and find almost any place in the world.

The Indiana Web Academy

www.indianawebacademy.org/students.asp

The Indiana Web Academy has created a large collection of websites that offer content and activities. As the site says, "Students, IWA has programs and resources designed to make your lives a bit easier." The site also includes an E-Locker section that provides students with online storage for school projects.

Eduhound

www.eduhound.com

Eduhound is a classic website for both teachers and students. As the website says, "Everything for education K-12. We tracked it down so you don't have to!" It supports a huge collection of links to just about everything related to schools. The science section is very easy to navigate.

Bottle Biology

www.bottlebiology.org

Check out this site for many ideas of how to do real science experiments and investigations using recyclable objects.

Science Learning Network

www.sln.org

If you are looking for access to science news, links to some of the best science museums, or inquiry-based science activities, this is the place. As the site states, "The Science Learning Network (SLN) is an online community of educators, students, schools, science museums and other institutions demonstrating a new model for inquiry science education. The project incorporates inquiry-based teaching approaches, telecomputing, collaboration among geographically dispersed teachers and classrooms, and internet/World Wide Web content resources."

Chemistry Teaching Resources

www.anachem.umu.se/cgi-bin/pointer. exe?GeneralScience

Even though the site "attempts to present a comprehensive list of chemistry teaching resources on the internet," it goes way beyond just chemistry in its general science section. The site is supported by Umea University in Umea, Sweden, and was created by Knut Irgum and Svante Åberg.

Bad Science

www.ems.psu.edu/~fraser/BadScience.html

Bad Science is maintained by Alistair B. Fraser, a professor retired from Pennsylvania State University. Dr. Fraser states that Bad Science is "an attempt to sensitize teachers and students to examples of the bad science often taught in schools, universities, and offered in popular articles and even textbooks."

Helping Your Child Learn Science

www.ed.gov/pubs/parents/Science/index.html

The book on this site "provides examples of a few simple activities that can be done with children. It is an introduction to the wealth of material in many other books available in libraries and bookstores. It might also inspire us to make up our own experiments to see why and how things turn out the way they do." Although written for parents, science teachers can also learn a great deal about effective science teaching and appropriate lessons.

Search Engines for the World Wide Web
Google
www.google.com
Noted for its ease in navigation, Google can search literally thousands of web resources in just seconds. It can search not only the internet but also local collections of files on a personal computer.

Ask Jeeves
www.ask.com
Ask Jeeves offers a handy little section called "Ask Jeeves for Kids." Children (and teachers) can ask questions and Jeeves, the electronic butler, will search out websites where answers can be found. This site is easy to use and an excellent resource for both science students and teachers.

In this chapter, we examined several nontraditional approaches to science instruction that are useful in advancing the scientific language and literacy of ELLs. First, we described how informal science experiences—including visits to museums, planetariums, and other nonclassroom-based community resources—help enrich ELLs' understanding of science.

We next described science experiences such as school-based science fairs and family science nights and how they enhance community and family involvement in science learning. Within the context of these events, non-native English speakers from all backgrounds and cultures can discover the excitement of scientific endeavor.

Third, we examined the role of developing technologies and their influence on ELLs. With the ever-increasing use of computer technology and the internet, students from across the globe have virtually endless opportunities to add to their scientific knowledge and understanding. Such technologies can have a tremendous impact upon second-language acquisition. "It's a small world" is taking on a new and more important meaning with the passing of each and every day.

References

ASCILITE. 2000. Webquest Resources. Available online at *www.ascilite.org.au/conferences/melbourne01/workshops/workshop9.html*.

Foundation for Family Science. 1999. In *Family Science*, eds. D. Heil, G. Amorose, A. Gurnee, and A. Harrison. Portland, OR: Foundation for Family Science.

Internet Usage Statistics. The big picture: World internet users and population stats. (n.d.) Retrieved December 1, 2004, from *www.internetworldstats.com/stats.htm*.

National Research Council (NRC). 1996. National Science Education Standards. Washington, DC: National Academy Press. Online version at *www.nap.edu/books/0309053269/html/index.html*.

National Research Council (NRC). 2000. *Inquiry and the National Science Education Standards:* A guide for teaching and learning. Washington, DC: National Academy of Sciences. Available online at *http://books.nap.edu/html/inquiry_addendum/ch2.html*.

National Science Teachers Association. 1999. July. Informal science education. Retrieved December 12, 2004, from *www.nsta.org/position*.

Phinney, M. 1996. Exploring the virtual world: Computers in the second language writing classroom. In *The power of CALL*, ed. M. Pennington, 137–152. Houston: Athelstan.

Phoenix Talent School. 2003. *ELL students, technology and standards*. Phoenix, OR. Available online at *www.phoenix.k12.or.us/stories/storyReader$171*.

Further Reading

Ebenezer, J. V., and J. E. Lau. 2003. *Science on the internet: A resource for K-12 teachers*. Upper Saddle River, NJ: Merrill Prentice Hall. The second edition of this handy, helpful book contains everything science teachers could want to know about finding reliable science websites on the internet, using the internet to create hands-on science activities, and building a science curriculum based on internet resources.

Feyten, C., M. Macy, J. Ducher, Y. Makoto, E. Park, B. Calandra, J. Meros, and M. Yoshii. 2002. *Teaching ESL/EFL with the internet: Catching the wave*. Upper Saddle River, NJ: Prentice Hall. This guide helps teachers use the internet effectively and creatively in their ESL/EFL (English as a second language/English as a foreign language) classrooms. Designed by scholars and teachers of second-language acquisition and educational technology, this user-friendly text will equip readers with the necessary pedagogical principles for using technology in the classroom.

National Science Teachers Association. 2003. *Science fairs plus: Reinventing an old favorite*. Arlington, VA: NSTA Press. This book, a collection of 20 articles from NSTA's member journals, explores all aspects of getting ready for a science fair. You'll learn how to help students pick their projects, understand what makes for fair judging, and create innovative alternatives, including festivals and expos. Several articles debate the pros and cons of sponsoring a science fair.

Robertson, W. C. 2001. *Community connections for science education: Building successful partnerships*. Arlington, VA: NSTA Press. This practical guide helps teachers work together with science educators in the community to avoid common pitfalls that interfere with the many science-learning opportunities that exist in museums, parks, and science centers. The book offers tips and strategies on selecting community resources for field trips, developing and evaluating educational materials, and even arranging transportation.

Wetzel, D. A. 2004. *Weaving the web into K–8 science*. Arlington, VA: NSTA Press. This booklet has a wealth of internet resources for both new and veteran teachers, plus tips on how to use technology in the classroom.

Section III

Lessons for Science and Language Learning

Designing Lessons: Inquiry Approach to Science Using the SIOP Model

Jana Echevarria and Alan Colburn

Sitting in his science class, Jaime is excited to get started looking at the ant farms. A curious boy, he enjoys exploring things in this class. But sometimes he feels a bit lost. The other students know names of the objects they look at and use, and they talk about things that Jaime sort of understands from observing, but he wants more explanation and the words to understand better. He also gets frustrated sometimes when he doesn't know the words to use with his friends to describe the new things they are observing.

Jaime is one of a number of English language learners in Mr. Giesen's first-grade class. Some were born in the United States but speak predominantly Spanish, while a few others, like Jaime, immigrated with their families only recently. Mr. Giesen is very kind and tries to make the students feel comfortable in his class. Often, he will let them work with partners or in small groups so that the students who speak English well can support the English language learners who have good ideas but don't always know how to express them. Jaime likes it when Mr. Giesen writes some of the words on the overhead projector because then he can refer to the words while he's discussing what he and his friends are looking at.

Window
Into the
Classroom

High-quality, hands-on science activities are an ideal medium for students learning English. When working with, manipulating, and thinking about things, a student often can understand concepts, even without the vocabulary to express the learning. Once students have experienced a new idea for themselves, the timing is perfect for the for-mal introduction of concepts that explain what they have now personally experienced.

In fact, the science education community firmly believes that quality experiences before the introduction of new concepts benefit all students, regardless of how many languages they speak (Lawson, Abraham, and Renner 1989). As we see in the "Window Into the Classroom," English language learners (ELLs) like Jaime enjoy hands-on experiences but need more explicit teaching of terms and concepts to make connections between what they have experienced and what the textbook and teacher say about the lesson's concepts.

We have organized an approach to teaching science to ELLs into three parts. First, we describe inquiry-based instruction and the learning cycle in the context of science lessons as well as the Sheltered Instruction Observation Protocol (SIOP) Model, an instructional approach that makes content material understandable for English language learners. Next, we present a lesson designed by a science educator followed by comments from an expert on ELL instruction. Finally, we describe a lesson designed by the ELL expert with comments by the science educator. The two perspectives—science and language—demonstrate the compatibility of the SIOP Model with the learning cycle.

Inquiry-Based Instruction

The interaction of science content and student learning can be represented in many different ways. The range of student engagement goes from passive observation to more active interaction, both physically and cognitively. The term *inquiry* represents a variety of classroom activities that are considered by most teachers as realistic, interactive, and engaging. Schwab (1966) categorized hands-on activities in terms of student involvement and the level of inquiry via responses to three questions, as shown in Figure 1.

In traditional school labs each of these questions has the same response: the teacher or text chooses the activity students will investigate, provides step-by-step directions

on what students should do, and tells students what to pay attention to and record during the activity. This type of lab, often called a "cookbook activity," is almost the same thing as a student doing a demonstration. Cookbook activities may be superior to alternatives if the teacher's major goal is for students to practice and learn particular laboratory skills. Cookbook activities can also be effective in helping students see an example of a particular scientific principle, like a demonstration.

For helping students learn to like science, think like scientists, and understand the nature of science, however, other types of lab activities in the hands of skilled teachers are often superior, as has been noted in the Chapter 2 discussion of best practices and in the Chapter 4 description of an inquiry continuum.

The opposite of cookbook activities are lab activities in which students do pretty much everything independently: they decide what to investigate and create their own investigations. These lab activities mimic the activities of science researchers and are probably best embodied in the classic science fair activity in which students investigate a question about which they are curious.

Advantageous as these types of activities are for giving students an experience like that of real science, including the opportunity to struggle and succeed with questions and data, there are reasons open inquiry laboratories are fairly rare in school science. They can be time-consuming, require more varied materials than typical school science activities, and require broader background knowledge than other activities, because students are engaged in different investigation. In addition, because open

Figure 1: Hands-on activity categories

Questions

> 1. Who decides the question students are investigating in the lab activity?

> 2. Who decides the procedures students follow to try answering the lab questions?

> 3. Who figures out the meaning of the data students create during the activity?

Types of instruction regarding questions 1–3

> **Traditional**
> 1. Teacher/text.
> 2. Teacher/text.
> 3. Teacher/text.

> **Structured Inquiry**
> 1. Teacher/text.
> 2. Teacher/text.
> 3. Students.

> **Open Inquiry**
> 1. Students themselves.
> 2. Students figure it out for themselves.
> 3. Students.

> **Guided Inquiry**
> 1. Teacher/text.
> 2. Students figure it out for themselves.
> 3. Students.

inquiry activities are different from typical school science activities, students often are unprepared for the challenges and teachers often are unprepared to deal with the challenges raised by students. Look first to other types of inquiry activities, and return to open inquiry activities when you and your students feel comfortable with open-endedness in the laboratory.

Structured and guided inquiry lie between open inquiry and cookbook activities. To best understand them, go back to Schwab's questions in Figure 1 (see Chapter 4 for an inquiry continuum). Structured inquiry labs are somewhat similar to cookbook labs. Where they differ lies in the third question, which is about decisions regarding data and their meaning. In structured inquiry activities, students make decisions about collecting data and about the data's meaning. Thus, you can create a structured inquiry lab from a traditional cookbook activity if you remove premade data tables or other items that tell students specifically what to observe. You might also remove some postlab questions related to direct observations and their meaning. In place of premade data tables, students are simply told to record the lab's important data.

Students accustomed to cookbook activities might take a while to get used to these changes. But before you know it, they will be comfortable without being told everything to record. Along the way, you have opportunities to discuss how scientists decide what is more or less important and to discuss various ways to record data. Which methods did students think worked well? What made it easy to understand the relevant data? Later in this chapter we provide guidance on making the transition from cookbook to more open-ended activities.

Guided inquiry lab activities are more open-ended than structured inquiry, but do not reach the level of open inquiry. Often, guided inquiry takes the form of the teacher presenting students with a question to investigate, providing relevant materials (and perhaps brief information about how to set up the materials), and turning students loose to figure out how to answer the question. For example, a teacher may provide students with string, washers, and a clock on the wall, and ask, "How can you make a pendulum that will swing back and forth 60 times in a minute?" Or for the classic "sinking and floating" unit, a teacher might ask students to shape a piece of clay so that it will float in water. In life sciences, teachers often give students mealworms and other materials, asking students to figure out whether the critters prefer light or dark surroundings. These are all guided-inquiry activities.

With these definitions laid out, you can see that structured, guided, and open inquiry activities represent a continuum of teacher and student behaviors. The borders between the different types of activities are not rigid.

The results of research on inquiry-based science activities and student test scores is mixed: inquiry as a teaching method can, but does not always, produce higher scores. Students engaged in inquiry tend to like science more and, to the extent we can assess these outcomes, improve their abilities to think and figure things out for themselves. Teachers can also affect students' conceptions of the nature of science by helping students see how their thoughts and actions mirror those of real scientists.

The Learning Cycle

The preceding section discussed inquiry-based instruction within the context of individual science activities. The learning cycle approach, on the other hand, is directed toward sequences of activities. As a science-teaching idea, the learning cycle is well established. Although the phrase *learning cycle* was not coined until the 1960s (Atkin and Karplus 1962), the approach was being advocated in the United States more than 100 years ago by the many followers of Herbart (DeBoer 1991) and numerous other educators. The basic concepts underlying the learning cycle approach are shown in Table 1.

Concepts related to electrical circuits provide an excellent example of a learning cycle approach in action. In the exploratory phase of the cycle, students are challenged to light a bulb when given only a battery, piece of wire, and the bulb. Only *after* most students have figured out various ways to light the bulb and have figured out rules for predicting whether or not a bulb will light are they ready to be formally introduced to the concept of an electrical circuit. After this concept introduction, students apply their new knowledge by trying to figure out how to light two bulbs, given all the wire they want. In this activity students apply and cement their newfound knowledge about electrical circuits while extending their learning to a more complex situation. The cycle will continue when the teacher goes on to discuss concepts of series and parallel circuits with students.

In recent years, the learning cycle idea has been expanded into instructional sequences like the "4E model" used in the Full Option Science System (FOSS) curriculum

Table 1: The learning cycle

- Lessons should start with students having a direct sensory experience (to the extent possible) with the new ideas they will eventually be expected to learn. This is a fancy way of repeating the experience-first idea. At this exploratory phase of the cycle, students are not being introduced to new vocabulary.

- Later, after students have had the chance to discuss their experiences and learning with the teacher and classmates, the teacher can then formally introduce the new concepts and terminology to students.

- Finally, maximum learning will occur if students then have the chance to use their learning in a new situation. Often called the application phase of the cycle, this is the time when students simultaneously cement and extend their learning.

developed by the Lawrence Hall of Science and the 5E model advocated by the Biological Sciences Curriculum Study (BSCS) that includes the steps of *engage, explore, explain* (i.e., concept introduction), *expand* (i.e., application), and *evaluate* (Bybee 2002).

Whichever learning cycle version you use, teaching English language learners requires specific attention to the unique language support needs of these students. The SIOP Model is an empirically validated approach for teaching English language learners that offers features that are effective in meeting these students' needs.

The SIOP Model

The SIOP Model (Echevarria, Vogt, and Short 2004) was developed during a 1996 to 2003 research project aimed at providing teachers with a well-articulated, practical model of sheltered instruction. The SIOP Model is used in nearly every state and in hundreds of schools across the United States, as well as in several other countries (see *www.siopinstitute.net*). The intent of the

model is to facilitate high-quality instruction for ELLs in content areas that include science, social studies, and math.

The SIOP Model of instruction for ELL is an ideal complement to inquiry-based science lessons. With an emphasis on providing meaningful, hands-on activities that provide English language learners with an opportunity to use academic English by interacting with peers, the model has features that increase the effectiveness of science lessons for ELLs. When students are learning new content at the same time they are acquiring a new language, certain supports are necessary for them to have a successful learning experience, even when an interesting lesson follows inquiry science teaching.

The SIOP Model comprises eight components with 30 features. The components reflect what we know about effective subject matter instruction, i.e., science teaching, as well as high-quality instruction for English language learners. The eight components are:

- *Preparation.* Good instruction is the result of good planning (Garcia and Beltran

2003). The preparation component's six features examine the lesson-planning process, including the incorporation of language and content objectives, the use of supplementary materials, and the meaningfulness of activities. Lesson objectives are critical for ensuring that students end up where you want them at the end of a lesson, unit, semester, or year. In planning for success for students, "Be clear on what students must know, understand and be able to do in order to grow in their grasp of a subject. Teacher fog will only obscure an already difficult view for struggling students." (Tomlinson 2001, p. 14).

- *Building background.* The importance of student background knowledge and experience in learning is well documented (Chiesi, Spilich and Voss 1979; Dole et al. 1991; Goldenberg 1992–3; Jimenez, Garcia, and Pearson 1996). This component focuses on making connections with students' background experiences and prior learning and developing their academic vocabulary.

- *Comprehensible input.* Coined by Stephen Krashen (1985), the term *comprehensible input* means making instruction understandable to students. The features of this component relate to adjusting teacher speech, modeling academic tasks, and using multimodal techniques to enhance comprehension.

- *Strategies.* The SIOP Model promotes explicit teaching of the strategies found to be used by proficient learners. Many ELLs have not had adequate academic preparation for their grade level and may not use strategies that successful learners tend to use, such as predicting, highlighting text, and using mnemonics for memorization. The strategies component emphasizes teaching students how to access and retain information. It includes scaffolding instruction and techniques for promoting higher-order thinking skills. When you model strategy use and then provide appropriate scaffolding while students are practicing the strategies, they are more likely to become effective strategy users (Fisher et al. 2002).

- *Interaction.* Research tells us that, in most classrooms, teacher speech dominates the setting, leaving students with limited opportunity to use language in a variety of ways (Goodlad 1984; Marshall 2000; Ramirez et al. 1991). Learning is more effective when students have an opportunity to participate fully by discussing ideas and information. Further, interaction among students and between teacher and students provides an opportunity for ELLs to practice using academic English. The interaction component of the SIOP Model is a reminder to encourage student participation by prompting elaborated responses, providing sufficient wait time for student response, and grouping students appropriately to meet language and content objectives.

- *Practice/Application.* This component calls for activities that extend language and content learning while integrating reading, writing, listening, and speaking into lessons.

- *Lesson delivery.* You should present a lesson that meets the planned objectives and keeps students engaged. Instruction that is understandable and meaningful to ELLs, that creates opportunities for students to talk about the lesson's concepts, and that provides hands-on activities to reinforce learning captures students' attention and keeps them more engaged. Inquiry lessons and the learning cycle both provide opportunities for students to be engaged in learning.

- *Review/Assessment.* The four items of this component consider whether you have reviewed the key language and content concepts, assessed student learning, and provided feedback to students on their output.

Let's take a look at one component, preparation, as an example of how it fits with inquiry-based science instruction.

The preparation component has six features:

1. Clearly defined content objectives, based on the National Science Education Standards (NRC 1996) or state standards.
2. Clearly defined language objectives, based on national TESOL (Teaching English to Speakers of Other Languages) or state English language development standards.
3. Content concepts appropriate for students at their grade level.
4. High use of supplementary materials including realia, pictures, and supplementary texts.
5. Content adapted to student's level of proficiency, both academic and English.
6. Meaningful activities that provide practice in reading, writing, speaking, and listening.

English learners are often left behind in academic areas, not only because they may not understand the science material but also because they aren't learning the academic vocabulary associated with the subject. They may get by in a lesson without fully learning the concepts and language needed for academic success, including success in standardized testing. In our work with English learners, we have observed classes where students are coloring pictures while the rest of the class is working with rigorous scientific concepts. Feature 3 implies that just because some students don't speak English fluently doesn't mean that they are limited in their ability to think about and learn from grade-level material. You should plan opportunities for ELLs to use academic language through reading, writing, listening, and speaking in meaningful ways as in Feature 6. Many ELLs hit a plateau in their level of English proficiency because they haven't had sufficient opportunity to use academic English meaningfully. They can speak the language but struggle with academic tasks, which require a different level of language usage. It is critical that you provide English learners with such practice within content area instruction.

Writing content and language objectives are intended to add clarity to the lesson rather than to disclose what the students are going to explore. English learners need as many contextual clues as possible to understand the material and the tasks they are expected to complete; objectives provide a road map for them. For example, if the class will be observing ant farms and the life cycle of ants and documenting their behaviors, the content objective would say something like, "To pose a question to be investigated and document the findings." The language objective might be, "To discuss observations and write your conclusions using complete sentences." As you can see, some words in the objectives may be unfamiliar to most ELLs, such as *pose, document,* and *conclusions.* These terms, which may need to be defined and discussed before the ant farms are introduced, are the very kinds of academic vocabulary words that we often assume students already possess.

In keeping with the learning cycle, after students have had an interesting, meaningful inquiry experience, you should write down vocabulary words that emerge during the discussion phase. It is important to write the words on an overhead projector, on science word walls, bulletin boards, or on the board as a reference for ELLs as the class continues to use these terms throughout the discussion and application phases. Add more terms to the vocabulary word bank as you formally introduce new concepts and terminology. With this kind of language support, English learners have a greater chance of extending and applying their learning in the application phase of the learning cycle.

The SIOP Model provides a guide for making sure that the features of effective instruction for English learners are incorporated systematically into content lessons. It is not a step-by-step process, but rather one that allows flexibility in how the features are addressed in a given lesson.

Merging Inquiry and the SIOP

Most readers are probably more knowledgeable about one of the instructional approaches—inquiry or the SIOP Model—than the other. To address this, one of the sample lessons was written by science educator Alan Colburn with commentary from ELL expert Jana Echevarria. The other example lesson was written by Jana Echevarria, using the SIOP Model lesson planning, with commentary from Alan Colburn. This approach should help readers with disparate viewpoints understand how inquiry and the SIOP Model can go together.

Some common misconceptions interfere with teachers' fully understanding how the ideas are complementary. First, the SIOP Model asks teachers to create for each lesson content objectives that are to be shared with students. These objectives need not be limited to factual or conceptual outcomes typically considered to be content. In other words, content is defined broadly; an inquiry-based lesson need not begin with discussion about ideas students will be learning for several days. A day's science objective may involve a process such as "Students will be able to observe and document behaviors of ants." An accompanying language objective might be "Students will be able to list three behaviors and report them to their group."

A lesson must begin with enough context to make it understandable for ELLs and must include the vocabulary necessary for participation in the activity. ELLs need to know what they are supposed to do and what you expect them to get out of the day's activity. Traditional content objectives, particular concepts or facts, can be stated explicitly at the (learning cycle's) concept introduction stage, *following* the exploratory activity.

A lesson can be defined in terms other than a block of time on a single day; it can take place over several days. This means, among other things, that science teachers need not address every single item on a SIOP checklist each day of the week.

A Lesson Designed by a Science Teacher: Science Lesson Perspective

To drive these important points home, this activity was chosen from another NSTA Press book, *The Truth About Science* (Kelsey and Steel 2001). The book is a hands-on curriculum designed to introduce middle-school-aged students to the processes of science research. Pages 16 and 17 discuss a lesson centered on a cornstarch and water mixture that will be familiar to some readers as "oobleck." The authors of this book call it "Ooze." Regardless of the name used, cornstarch mixed with just the right amount of water produces a thick, viscous material that has properties of both liquids and solids. If you press slowly, your finger will go through the substance as it would any thick liquid. If you hit it with a hammer, however, you will find it cracks—as do solids.

As a science educator, I chose this activity to comment upon for two reasons: it represents one day's activities, but is only part of the overall lesson; and it is less content focused than many science lessons.

As set up in *The Truth About Science*, students had previously explored Ooze's properties and come up with questions about the stuff to investigate. The student investigations ultimately represent guided or perhaps even open inquiry activities. In the previous day's activities, students had written a procedure to use to answer their question via an investigation. The current lesson should be centered on the focus question of "How do you report the results of scientific research?'

With materials ready and the classroom prepared for messiness, the day's activity has several parts. First, students make Ooze and carry out their research plans (which need to include quantitative measurements). After finishing and cleaning up, students write about their results and conclusions. This is followed by whole-class discussion about student results. This whole-class part of the day's activities includes reminding students about report requirements that had been previously discussed, such as the research question, methods, results, and discussion. Notes from the book add:

> If there is time, you might want to introduce the concepts of treatment types and replication. Define the concepts and ask students how controls, alternative treatments, and/or replication might have been useful for their experiments. These concepts will be formally introduced [later] (p. 17).

Your role during the hands-on portion of the lesson would be managing the classroom setting, including the various materials that students were using. In addition, you would watch students and talk with them to better understand how they were interpreting their work and to keep them focused on the lesson's key task(s). You might say

- "What are you observing?"
- "Tell me about what you are thinking."
- "What did you think would happen?" and "Why did you think so?"

To keep students focused, ask: "Has your procedure ended up being a little different than you'd originally planned?" You could follow by saying, "Don't forget to write down the differences."

With this description as a starting point we now turn to the ELL expert for commentary and suggested changes to the lesson when viewed via the SIOP Model.

First, begin with a brief review of the previous day's lesson. Students often need an explicit linking of previous work to the activity at hand. In fact, I think it would be preferable to review the report requirements—the research question, methods, results, and discussion— to set the context. Then discuss the objectives—content and language. It is easiest to write good objectives by using the acronym, SWBAT (Students Will Be Able To ...) as a guide. That way, you can more easily assess student learning because you are clear about the direction of the learning. For this lesson the content objective would be something like: students will be able to tell how to report the results of scientific research, and students will draw one or more conclusions about their research. The language objective might be: students will report their findings using complete sentences and correct punctuation.

After you go over the objectives for the day, the students would begin carrying out their research plans. Grouping is important at this point (interaction) so that native English speakers or more proficient students are paired with ELLs who lack the language skills to complete the activity. ELLs may need support from the teachers and/or peers to express their learning, take notes, or complete the report (consistent use of scaffolding techniques in strategies). Further, partnering with a proficient speaker provides the ELL student opportunities to hear and use academic English (interaction) that will be needed to complete the report. Some ELL students may need an outline of a report or a graphic organizer to complete in lieu of writing a paragraph or more (adaptation of content for proficiency level

in preparation).

The teacher questions suggested for the hands-on phase are appropriate, and you can assist the ELL student by modeling or paraphrasing as needed. For example, say, "Tell me about what you are thinking," and the student might reply, "Think it needs water to be soft." You would scaffold by saying something like, "Oh, so you think the Ooze needs water to remain soft and not become hard. OK, let's write that down." Then, you or the student's partner could assist in making sure the student writes a complete sentence.

During the final phase, the class discussion of results, you would use an overhead projector or other display device to write key vocabulary words and make sure the students know their meaning. Further, you would jot down vocabulary as it comes up, model writing of complete sentences as students report their results, and write or draw words and illustrations as often as possible to provide the ELL students with contextual clues as referents during the discussion.

A Lesson Designed by an ELL Expert: Language Lesson Perspective

To illustrate the SIOP Model, use a lesson about ants that relates to the study of living things, a common unit covered in first grade. Begin with the objectives. You would read from the board, "Today we will explore the lives of ants. You will tell one question you have about ants after you have watched them." This introduction guides students' exploration and lets them know what is expected; it doesn't tell them what they will observe.

For young children, express the content and language objectives in terms that they understand. Your plan book would say, "SWBAT name one question about the life of an ant" for the content objective and "SWBAT use key vocabulary to discuss lives of ants" for the language objective.

Then pass out an ant farm to each group of four students. The students observe the ants and come up with questions they'd like answered about the lives of ants. Group English learners with more proficient speakers to help the ELLs express their ideas more fully (*interaction*). Circulate around the classroom and assist students as needed, asking questions like those in the previous lesson:

- What do you notice about the ants?
- Why do you think they do that?
- What would you like to know about the lives of ants? Why?

After students have discussed their questions in their groups, tell them to write at least one question on their paper (*practice/application*). Then, a whole-group discussion ensues. Be careful to elicit key vocabulary from the students and write the terms on the board (or overhead) such as *ant farm, insect, legs, queen,* and *carry*. (Key vocabulary is emphasized in *building background*.) Finally, read the objectives from the board and ask the students if they met the day's objectives (*review/assessment*). Conclude by telling students that the next day they will begin exploring answers to the questions they posed.

Comments on the lesson by the science educator:

I will assume that students are observing a mature ant farm, tunnels are visible, and ants are active throughout the habitat. From my perspective as a science educator, a lesson like this has a lot of potential. The key points, in this case, relate not to the materials or instructions, which *are fine, but to the kinds of questions the teacher asks and how she responds to students' ideas. Sometimes young children have rich knowledge about ants, thanks to having observed ants in and around their homes and on television programs. Asking young children about things they know related to ants could elicit an interesting list you can use to focus student observations while observing the ant farm.*

Students will likely observe ants eating, carrying dead ants, and carrying other objects much bigger and heavier than themselves. Student observations can be focused toward similarities among ants, humans, and all living things via questions like

"What do you see the ants doing that you also do?"

"Do you see any ants that you think are sleeping?" "Can you show me?"

"Do you see any ants that are eating?" "Can you show me?" "Why do you think they are eating?"

The last question could lead toward a class experiment in which students try to discover the kinds of foods that ants do and don't like to eat. My state's grade–1 life science standards include an item about animals' basic needs and that they inhabit different kinds of environments. So the trick with this lesson relates to listening to students' questions, responding in ways that encourage students to think and tell you more, and basically waiting until a student brings up something about food (or shelter or some other basic need). As the teacher, you are waiting and gently leading students toward your (already-planned) next lesson, in which they will investigate foods that ants do and do not eat (or some other investigation you feel comfortable leading). Student observation, followed by a lesson on what ants eat (and the fact that all living things need to eat), could easily represent the exploratory and content introduction phases of a learning cycle. The investigation that follows would be structured or guided inquiry, depending on how much guidance your first graders need in creating a fair test.

A further lesson might be one in which students go outside and compare ant behaviors on the playground, or near their homes, with the behaviors they see in the ant colony. You can help students imagine what the underground ant colonies look like, with students applying their knowledge of ants in ant farm colonies to those of "real" ants in the wild.

Caution from ELL expert:

I really like the suggestions for deepening students' understanding of ants as living things. All the lesson ideas mentioned are consistent with the SIOP Model and getting students to think about and understand new concepts and ideas in meaningful ways. You as a teacher, however, would need to be mindful of posting new words used during the discussion, checking for understanding frequently, and scaffolding ELL students' contributions. For example, a beginning speaker may only be able to point and say, "Not eat." In this case, the teacher would say, "OK, Huong, you don't see the ant eating. Does anyone else see an ant eating?" This kind of verbal scaffolding models complete sentences for ELLs while empowering these students to contribute to the lesson at their level of English proficiency.

Conclusion

Once understood, inquiry-based science lessons and models like SIOP make for a natural and powerful combination that enhances all students learning. Science lessons are most interesting and meaningful to students when they involve inquiry, allowing students to act like scientists by exploring questions of interest. For ELLs, learning is enhanced when teachers use techniques that make the information and ideas understandable for them, regardless of their level of English proficiency. ELLs can think about important ideas even though they may not be able to fully express them. Your role is to create a classroom environment in which these students, and all students, feel accepted, encouraged, and even empowered to participate in learning.

References

Atkin, J., and R. Karplus. 1962. Discovery or invention? *The Science Teacher* 29 (5): 45–51.

Bybee, R., ed. 2002. *Learning science and the science of learning.* Arlington, VA: NSTA Press.

Chiesi, H., G. Spilich, and J. Voss. 1979. Acquisition of domain-related information in relation to high- and low-domain knowledge. *Journal of Verbal Learning and Verbal Behavior* 18: 257–274.

DeBoer, G. 1991. *A history of ideas in science education. Implications for practice.* New York: Teachers College Press.

Dole, J., G. Duffey, L. Roehler, and P. Pearson. 1991. Moving from the old to the new: Research in reading comprehension instruction. *Review of Educational Research* 61:239–264.

Echevarria, J., M. Vogt, and D. Short. 2004. *Making content comprehensible for English learners: The SIOP model,* 2nd ed. Boston: Pearson Allyn and Bacon.

Fisher, D., N. Frey, and D. Williams. 2002. Seven literacy strategies that work. *Educational Leadership* 60: 70–73.

Garcia, G., and D. Beltran. 2003. Revisioning the blueprint: Building for the academic success of

English learners. In *English learners: reaching the highest level of English literacy*, ed. G. Garcia, 197–226. Newark, DE: International Reading Association.

Goldenberg, C. 1992–93. Instructional conversations: Promoting comprehension through discussion. *The Reading Teacher* 46: 316–326.

Goodlad, J. 1984. *A place called school: Prospects for the future.* New York: McGraw-Hill.

Jimenez, R., G. Garcia, and P. Pearson. 1996. The reading strategies of bilingual Latina/o students who are successful English readers: Opportunities and obstacles. *Reading Research Quarterly* 31: 90–112.

Kelsey, K., and A. Steel. 2001. *The truth about science: A curriculum for developing young scientists.* Arlington, VA: NSTA Press.

Krashen, S. 1985. *The input hypothesis: Issues and implications.* New York: Longman.

Lawson, A., M. Abraham, and J. Renner. 1989. A theory of instruction: Using the learning cycle to teach science concepts and thinking skills (Monograph Number One). Kansas State University, Manhattan, KS: National Association for Research in Science Teaching.

Marshall, J. 2000. Research on response to literature. In *Handbook of reading research*, Vol. 3., ed. M. L. Kamil, P. B. Mosenthal, P. D. Pearson, and R. Barr, 381–402. Mahwah, NJ: Lawrence Erlbaum.

National Research Council. 1996. *National science education standards.* Washington, DC: National Academy Press. Online version at *www.nap.edu/books/0309053269/html/index.html.*

Ramirez, J., S. Yues, D. Ramey, and D. Pasta. 1991. *Executive summary: Final report: Longitudinal study of structured immersion strategy, early-exit and late-exit transitional bilingual education programs for language minority children.* (Contract No. 300-87-0156) Submitted to the U.S. Department of Education. San Mateo: Aguirre International.

Schwab, J. 1966. *The teaching of science.* Cambridge, MA: Harvard University Press.

Tomlinson, C.A. 2001. *How to differentiate instruction in mixed-ability classrooms.* Alexandria, VA: Association for Supervision and Curriculum Development.

Further Reading

Colburn, A., and Clough, M. 1997. Implementing the learning cycle. *The Science Teacher* 64 (5): 30–33. Research supports the learning cycle as an effective way to help students enjoy science, understand content, and apply scientific processes and concepts to authentic situations. The learning cycle is one technique teachers can use to diagnose and change students' conceptions about scientific principles. The authors discuss how a teacher can gradually move from traditional teaching to a learning cycle based approach.

Colburn, A. 1997. How to make laboratory activities more open-ended. *CSTA Journal* (Fall) 4–6. Available at *www.exploratorium.edu/IFI/resources/workshops/lab_activities.html.* This article presents ideas about how a teacher can gradually adopt inquiry-based approaches, making the transition slowly and gently. In-depth knowledge and examples show how a teacher can make changes in small steps or in great leaps and bounds.

Echevarria, J., and A. Graves 2003. *Sheltered content instruction: Teaching English language learners with diverse abilities*, 2nd ed. Boston: Allyn and Bacon. In this book, sheltered instruction, or specially deigned academic instruction in English, is clearly defined and strategies for its successful implementation in the classroom are provided. Focusing on the use of sheltered instruction with students of varying abilities, the text addresses the important overlap between sheltered instruction and special education adaptations.

Echevarria, J., M. E. Vogt, and D. Short. 2004. *Making content comprehensible for English learners: The SIOP model*, 2nd ed. Boston: Allyn and Bacon.

This text presents an empirically validated model of sheltered instruction and contains the Sheltered Instruction Observation Protocol (SIOP). The protocol provides school administrators, staff developers, teachers, university faculty, and field experience supervisors with an instrument for observing and quantifying a teacher's implementation of quality instruction for English language learners. The items on the protocol combine to constitute the SIOP Model of instruction, and, within the text, each item is illustrated with descriptions of the model in practice across a variety of subject areas and grade levels.

Lessons That Work: Science Lessons for English Learners

Ann K. Fathman and Olga Amaral

Ms. Smith used to begin her unit on soils by having students read from a science textbook and by explaining to them that there are many different types of soils. She followed this with worksheets and more presentations about causes for various types of soils. Today, Ms. Smith has a classroom of 20 students, 17 of them English learners. Five have recently arrived from Mexico and speak only Spanish. One is the daughter of a Korean family working in Mexico but living in the United States.

Ms. Smith has learned to modify lessons to accommodate the diversity in her classroom. A visit to her classroom today shows a different instructional pattern. First, Ms. Smith begins the lesson by holding up some of the materials that students will use, and students call out the name of each. For example, the class will use wire mesh to separate items used in the exploration phase of the lesson. Once a student correctly identifies the wire mesh, Ms. Smith asks the entire class to repeat the words and then hangs the item from a bulletin board next to a previously printed sentence strip with the words wire mesh. This stays on the bulletin board for the entire unit, and students can at any time look at the board to get linguistic cues and help with spelling.

Window *Into the* Classroom

The students will explore for themselves different types of soils. They will work in pairs (one beginner or intermediate English learner paired with a student who has more proficiency) to make observations and then to write focus questions, based on their observations, about what more they'd like to know about soils.

Ms. Smith leads students through this investigation with questions of her own. She has carefully planned to get students to chart information about their findings during their explorations. Ms. Smith finds that writing their findings on a chart on the board helps all her students organize information, a critical skill for future science

Window
Into the
Classroom

lessons. Finally, her students write sentences with the assistance of sentence starters that help them record what they have learned during this lesson. Ms. Smith is pleased that even her most recently arrived students participate and that she has seen that peer assistance during the discovery portion of the lesson.

Ms. Smith's lesson is but one of many that have been identified as having essential components of both science and language learning practices that are particularly appropriate for English learners. This chapter provides examples of lesson plan formats and 13 sample lessons that have been used successfully to teach science to English language learners. A discussion of successful lesson components follows the lessons.

Sample Lesson Plan Formats

Teachers can use many lesson plan formats when developing science lessons for English language learners (ELLs). Terms to describe activities and focus in lessons may vary, but the sequencing of activities and certain basic components should be included. The most effective lessons are those that are inquiry-based and include components that make information and ideas understandable to English language learners, such as the strategies described in Chapter 5. Some options for describing the teaching phase in lesson plans using inquiry formats are

- preview, presentation, practice, assessment (generic inquiry);
- exploration, concept development, application (learning cycle); and
- engagement, exploration, explanation,

elaboration, and evaluation (5E model learning cycle).

Other possible formats for ELL lessons that include inquiry-based activities but use different terminology in describing activities in the teaching phase include

- motivation, preparation, practice, application, review/assessment, extension (Echevarria et al. 2004);
- introduction/background/motivation, guided instruction, independent activity (Shin 2005);
- teacher demonstration, group activity, independent activity—based upon a continuum of inquiry activities (Fathman and Quinn 1989); and
- setting the stage—preparation, providing input-presentation, guided participation-practice, evaluation, extension (Haley and Austin 2004).

The sequencing and type of activities are what is important—not the terminology used to describe the activity in the lesson plan. A variety of publications are available that describe science lessons designed for English learners (see, for example, Bassano and Christison 1992; CISC 2002; Hill, Little, and Sims 2004; Shin 2005). The majority of work in lesson development is being done by individual teachers who teach science in their mainstream, science, bilingual, or English language development classes.

Sample Lessons

The lessons that follow share some common elements that make them beneficial to English learners. All of them have components that activate student prior knowledge, an essential element for helping English learners begin to understand the context for the lesson. They also have inquiry-based activities, which provide English learners with opportunities to participate in authentic experiences. A detailed discussion of these similarities follows at the end of the chapter. Each lesson presents activities that will lead students to develop conceptual understanding based on content standards, either state or national, that are appropriate for a particular grade level.

Language objectives or standards also accompany each lesson. These are designed to work harmoniously with the science content standards to introduce students to aspects of English language development while they study science content. Within this framework, the lessons present a cycle of exploration, concept development, and application designed to get students to complete a learning cycle in each topic.

The lessons are grouped by grade: for the primary grades, K–2; for intermediate grades, 3–5; for middle school, 6–8; and for high school, 9–12. The lessons cover topics in life, physical, and Earth science. Each lesson includes a description of the topic, the school or class where the lesson was taught, background information on what the students need to know and what a teacher should need to know to teach the lesson, objectives or standards for science and language, materials needed, teaching activities, and a summary of why the teacher felt this lesson was effective.

Plants

Adhan J. Perez, Texas Avenue School, Atlantic City, New Jersey

Topic

Needs and characteristics of trees

Context

This lesson was taught in a kindergarten transitional class that received ESL support. There were 17 students in the class, representing China, the Dominican Republic, Honduras, Mexico, Pakistan, Puerto Rico, the United States, and Vietnam. Students were beginning to advanced English learners.

Objectives
Content Objectives
Students will be able to

Lesson

1

Kindergarten

- identify needs of a plant
- estimate age of a tree
- model being a tree

Language Objectives

Students will be able to

- listen to the story *The Giving Tree* by Shel Silverstein (Harper Collins 1964) and respond to questions
- use the strategy "get your mouth ready to make the sound of the first letter" to read words aloud

Background

This lesson builds on the students' prior knowledge about the needs of plants and trees. The unit from *On Our Way to English* (Freeman et al. 2003) focuses on growth, both human and plant. This lesson was developed using the Sheltered Instruction Observation Protocol Modal (Echevarria, Vogt, and Short 2004) and supplemental materials.

Materials

picture cards	plates
poker chips	chart paper
marker	real tree trunk (use firewood or a stump)
seeds	book *The Giving Tree*
Reading strategy card #7 from *On Our Way to English*.	

Vocabulary to estimate, roots, leaves, minerals, sunlight

Motivation

Arrange each student with a partner in the read-aloud area. Tell the students they will be learning about trees. Using a K-W-L[1] (see end notes for this chapter) chart to organize information, ask students to identify what they know about plants and trees and what they want to know about plants and trees.

Show drawings from *The Giving Tree*, by Shel Silverstein, to activate prior knowledge. Use "buddy buzz," a think-pair-share[2] strategy, by asking students, "What kind of tree do you think this is? Are there any clues to help us guess what kind of tree this is? Could it be an orange tree, or a banana tree?" Have your students think about it, then buddy buzz, or share, their guesses with a partner.

Tell students that *The Giving Tree* is about two good friends, a boy and a tree. Using the strategy card example, model the strategy by saying "get your mouth ready for the first sound." This card has a picture of a boy looking at a book with the sentence "I see the fish." The boy is staring at the first letter in *fish*. There are four f's in front of his lip modeling the get-your-mouth-ready strategy. Explain to the children that they will use this strategy in different parts of the book when they come across some unknown words. Using the title and illustration, ask the students to predict some things a giving tree could give to a boy.

Presentation of the story:
Set the purpose for reading by telling students, "I am going to read to find out what a tree can give to a boy."

Read the story aloud, and stop to apply the sound-letter strategy on key words relevant to the content and the sound-letter identification curriculum. Stop periodically to ask students to recall what the tree has given to the boy and to predict what the tree will give next.

Provide students with the meaning of vocabulary they might not know. For example, ask, "What is the trunk of a tree?" If no one can answer, show them and say, "It is the part between the roots and the branches that supports the branches." Review or teach other vocabulary: *root, leaves, sunlight, minerals, water*. Refer to the K-W-L chart to help guide the vocabulary development from the Shel Silverstein drawings and the story itself.

Science Practice

This activity will help demonstrate how a tree uses its parts to acquire water, vitamins/minerals, and sunlight. Show students a cross section of a real tree trunk (use fire wood or a stump). Explain to the students that each ring represents a year's growth. Fat rings show a good year. Then, distribute a paper plate with rings drawn on them (to represent a cut tree trunk) to each student. Ask students to place their plates on the floor. Ask them to stand on their plates and remain stationary until the end of the activity.

Follow these steps:
1. Tell students the red chips represent sunlight; the blue, water; and the white, vitamins/minerals.
2. Randomly drop red, white, and blue poker chips on the floor around students.
3. Stand on one plate and model how students are to use one foot to collect as many of the poker chips as possible.
4. As you model this, remind students they can only use one foot, while the other remains stationary on top of the plate.
5. Demonstrate to students a start-stop signal for the beginning and the end of the collection activity.

Lesson

1

Kindergarten

6. Signal and start playing the game.
7. After most of the chips have been collected, signal a second time for the students to stop.

Application

Ask students which color they collected the "most" of and the "least." Model how you would estimate this with your own pile of colored chips. Explain the word *estimate*. Ask students to estimate how many chips they have of each color. Have students share their estimations aloud. Use a T-chart to record estimations on the left and actual numbers on the right. To get the actual numbers, allow some time for the children to count.

Encourage the children to share orally with sentence starters: "I used my roots ..." "I used my leaves to get ..." "I collected ..."

Ask discussion questions for a think-pair-share. "Which do you think is better for the plant to grow? To get a lot of red chips or sunlight? ...blue chips or water? ...white chips or vitamins/minerals? Do you think it's better to get a few of each colored chip? Think about it and share with a partner. What will happen to you, the plant, if you don't have any blue chips?"

Review/Assessment

Review all the objectives. The students will help determine whether or not the objective was achieved. Sketch a tree on the board and have students identify parts of the tree and tell how each part helps the tree grow. Fill in the last column of the K-W-L chart about what the class has learned as a shared writing review activity.

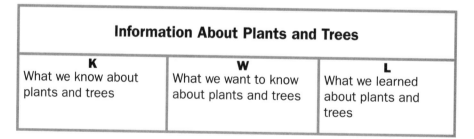

Information About Plants and Trees		
K What we know about plants and trees	**W** What we want to know about plants and trees	**L** What we learned about plants and trees

Summary

This lesson provides a very good example of how the integration of science and reading engages students, develops vocabulary, and improves reading skills. It also incorporates the use of prior knowledge and encourages student participation in recording of information and assessment.

Classifying Physical Characteristics

Lisa Mena, Heber School District, California

Lesson

2

Kindergarten

Topic
Myself and others

Context
This lesson was taught in a kindergarten class with 20 Hispanic students, 19 of whom were English learners.

Standards and Objectives

Science Content
This lesson partially meets California science standards in Life Science for Kindergarten (California Department of Education 2003, p. 1) Specifically it meets Standard 2 a,c as follows:

- Different types of plants and animals inhabit the Earth. As a basis for understanding this concept:
  - Students know how to observe and describe similarities and differences in appearance and behavior of plants and animals (e.g., seed-bearing plants, birds, fish, and insects), and
  - Students know how to identify major structures of common plants and animals (stems, leaves, roots, arms, wings, and legs).

Investigation and Experimentation
- Compare and sort common objects by one physical attribute (color, shape, texture, size, weight).
- Communicate observations orally and through drawings.

English Language Development
- Develop listening and speaking skills using vocabulary related to the body.
- Write simple words related to content area unit.

Background
This lesson is part of a larger unit in science designed for kindergarten students to understand systems. A unifying theme is that each of us is unique but we share some characteristics. Students need to begin to classify at this level.

Lesson

Kindergarten

Materials

Large roll of paper for silhouette cutouts

Activities

Engagement

The teacher asks students to comment on what is special about them and then records their answers.

Exploration

- Students observe the teacher modeling the design and formation necessary to form a life-size silhouette of a student on a piece of paper. The teacher uses one student as a model to lie down on the paper as she then traces the student's outline in pencil. She then cuts the form and asks students to do their own.
- The teacher asks students to work in pairs taking turns to model their silhouettes.
- The teacher asks students to observe their own silhouettes and think about what they look like.

Explanation

- Students reflect upon their own silhouettes, and the teacher explains that they are not all exactly alike, yet there are many similarities.
- The teacher elicits comments regarding items that they found to be alike and records these followed by those items that are different. These are recorded on a chart. Some of the vocabulary used is as follows:

Differences	Similarities
Height	Heads
Size	Hands
Hair	Feet
Shoes	Arms
	Legs
	Shoulders

Elaboration

The students copy the chart into their science notebooks. They can also draw the item each word represents to help them remember what the word is. The teacher encourages them to add to the chart if they can think of other differences or similarities.

Lesson

2

Kindergarten

Evaluation

The teacher asks questions about the chart and about their own silhouettes. The science notebook entries for the day can also serve as an assessment tool to determine students' level of completion of the task as well as whether they were able to go beyond what the teacher did during the group activity.

Summary

This lesson effectively integrates science and writing by having students record data based on observations.

Lesson

3

Grade 2

Soils

Victor Machuca, Blanche Charles Elementary School, Calexico, California

Topic

Soils: Introducing sand, clay, and humus

Context

This lesson was taught to a group of 20 Hispanic students in grade 2, 17 of them English learners at various stages of proficiency.

Standards and Objectives

Science Content

This lesson partially meets California science standards in Life Science for Grade 2 (California Department of Education 2003, p. 6). Specifically it meets Standard 3c, as follows:

- Students know that soil is made partly from weathered rock and partly from organic materials and that soils differ in their color, texture, capacity to retain water, and ability to support the growth of many kinds of plants.

Investigation and Experimentation

Scientific progress is made by asking meaningful questions and conducting careful investigations. As a basis for understanding this concept and addressing the content in the other strands, students should develop their own questions and perform investigations. Students will make predictions based on observed patterns and not random guesses.

Lesson 3
Grade 2

English Language Development
- Ask and answer simple instructional questions using simple sentences.
- Produce independent writing using standard word order, grammar forms, capitalization, and correct spelling (writing in the content area).

Background

Introduce at the beginning of a unit on soils. This follows a lesson in which the teacher determined prior knowledge by asking many questions about students' knowledge of soils and conducted an exploration of sample soils in the gardens or grounds of the school. This sets the stage for getting students to understand the unifying key concept or big idea, which is: soil is made up of different components, each of which has unique identifiable properties.

Materials

paper for charts	humus
clay	magnifying glasses
sand	cups and covers for cups
student science notebooks	

Activities

Engagement

The teacher arranges the students in small groups and gives each group three different types of soils. The teacher asks questions about what students might know about soils. For example, Do all soils look the same? Are they the same color? Then students place one type of soil in a cup. They listen to see what it sounds like when they shake it.

Exploration

The teacher asks questions about what the different soil types sound like. For gravel, expect answers like maracas, salt in a shaker, and a rattlesnake.

On a large chart with names of soils on the left column and the senses across the top, the teacher records responses as in the following example:

	Sounds like	Looks like	Smells like	Feels like
sand	maracas	gold	banana	rough
clay	pebbles	dirt, mud	nut	not too rough
humus	soil and sand	coffee	black crayon	moist

Students record the same information on their own charts while the teacher models. The teacher follows the same procedure for each type of soil and guides students with questions to get them to determine similarities or differences. The teacher records collective findings on the board and students record their own.

For the third type of soil, the humus, the students pair up and record their findings by themselves without the teacher's support.

Explanation
The teacher refers students to the chart in front of the class and asks students what can be said about soils. The teacher leads students through questions to get them to respond that soils are different and have different properties.

Elaboration
Students go on to the next page in their science notebooks and record what they learned about soils during the lesson. This activity is done individually.

Evaluation
- The assessment is informal with a review of the student responses and questions during the lesson.
- A review of student writing in science notebooks also informs the teacher of student progress.

Summary
This lesson is a good example of scaffolding instruction for English learners and vocabulary enhancement for various levels of English proficiency. It also demonstrates how to make meaning from observations and findings.

Life Cycles of Organisms

Angelica Oliva, Ben Hulse Elementary School, Imperial, California

Lesson

4

Grade 3

Topic
Life cycles of organisms (brine shrimp)

Lesson 4

Grade 3

Context

This lesson is the third in a series of 12 for a group of 19 students in grade 3. Fourteen were Hispanic, one Asian, and four White. Seven were English learners at various stages of English proficiency.

Standards and Objectives

Science Content

This lesson partially meets California science standards in Life Science for Grade 3 (California Department of Education 2003, p. 9). Specifically it meets Standard 3, as follows:

- Adaptations in physical structure or behavior may improve organisms' chance for survival. As a basis for understanding this concept, students know
  - Plants and animals have structures that serve different functions in growth, survival, and reproduction.
  - When the environment changes, some plants and animals survive and reproduce, and others die or move to new locations.
  - Living things cause changes in the environment where they live; some of these changes are detrimental to the organism or other organisms, whereas others are beneficial.
  - Some kinds of organisms that once lived on Earth have completely disappeared; some of this resemble others that are alive today.

Investigation and Experimentation

Differentiate evidence from opinion, and know that scientists do not rely on claims or conclusions unless they are backed by observations that can be confirmed.

English Language Development

Write an increasing number of words and simple sentences appropriate for language arts and other content areas (science).

Background

This lesson, part of a unit on the life cycles of organisms, uses brine shrimp as the example for students to explore. The overarching concept of the unit is to convey to students that organisms have specific structures that enable them to survive in their environment. This lesson is set within the context of a subconcept from which students learn that brine shrimp hatch from eggs. In the previous lesson, the teacher has introduced a K-W-L chart. For homework, students asked someone about resources they might use to find scientific information.

Materials

chart with pictures of various animals
container with brine shrimp eggs
a piece of white paper
magnifying lenses
student science notebooks
examples of science resources such as encyclopedias, videos, and dictionaries

Activities

Engagement

The teacher proposes to students that the focus question for this lesson is "Eggs? What kind of eggs?" and tells students that the lesson is about gathering data and identifying resources to gain information. They will also be conducting observations.

Exploration

Using sentence starters, the teacher guides students to list their predictions about the brine shrimp eggs, such as *"I think that it is* an egg of a fish *because* fish have little, tiny eggs" and *"I think that it cannot be* a fish egg *because* the fish egg is bigger." Many students contribute their predictions.

The teacher refers students to a previously prepared K-W-L chart that lists some information about eggs and discusses why a K-W-L chart is important (it helps with data collection). The teacher should expect questions like "Do mammals lay eggs?" The teacher reviews questions from the K-W-L with students and leads students to others such as "Can these be the eggs of a butterfly?" and "Does the butterfly lay its eggs in water?"

The teacher asks about resources students can use to find answers to some of their questions about eggs (their homework had been to ask someone about places in which one could find information) and writes their suggestions—such as ask parents, get real eggs, go to an encyclopedia or other books and videos—on the board.

The teacher discusses with students the difference between getting information from books of fiction and nonfiction materials.

The teacher distributes containers of solution with brine shrimp eggs and asks students to observe the eggs and record all of their observations in their science notebooks, first based on naked-eye observations and then on observations made with a magnifying lens. The teacher walks around the classroom providing encouragement and asking students what they notice, sometimes prompting them with questions such as "Do the eggs float?" and "Have they sunk or dropped to the bottom?"

Lesson 4
Grade 3

Explanation

The teacher reminds the students about the importance of observation and how they will be accessing information from various resources during a forthcoming lesson to see if their observations can be validated by what they already know about these eggs.

Elaboration

Students record their findings about their observations as the teacher circulates to ensure that they are including appropriate items in their science notebooks.

Evaluation

The assessment is informal, with a review of the student responses and questions during the lesson and then a review of student writing in science notebooks.

Summary

This lesson is a good example of how to help students focus on the process of acquiring knowledge in science, not just through the inquiry but also by gathering information by using reference materials of various types.

Lesson 5
Grades 3–5

The Sun
Fay Shin, California State University, Long Beach, California

These lessons have been used in fourth- and fifth-grade classrooms in the following school districts:

Temple City School District, Temple City, California
Stockton Unified School District, California
Los Angeles Unified School District, California

Topic
The Sun

Context

This lesson can be used for English language learners in grades 3 through 5. The lesson has been designed for students at the following three English proficiency level(s): beginning, early intermediate, and intermediate.

122

Standards and Objectives

Students will be able to describe and learn about the Sun and how it affects life on Earth. Students will also know the Sun is the central and largest body in the solar system and is composed primarily of hydrogen and helium. This lesson is part of a series of science lessons from *Journeys—ELD/ELA in the Content Area: Science* (Shin 2005).

California English Language Development Standards (California Department of Education 1997):

- ELD Level 1 Beginning *3–5*: Writing Strategies and Applications: Create simple sentences or phrases with some assistance.
- ELD Level 2 Early Intermediate *3–5*: Reading Comprehension: Read text and orally identify the main ideas by using simple sentences and drawing inferences about the text.
- ELD Level 3 Intermediate *3–5*: Reading Comprehension: Read text and orally identify the main ideas and details of informational materials, literary text, and text in content areas by using simple sentences.

California Science Content Standards (California Department of Education 2000):

- Third Grade—Earth Science: Energy and matter have multiple forms and can be changed from one form to another.
- Fifth Grade—Earth Science: The solar system consists of planets and other bodies that orbit the Sun in predictable paths.

Background

This lesson is part of a unit on the solar system. The students have just completed a lesson that introduced and gave an overview of the solar system. Student will know the Sun, Earth, eight other planets, many moons, and other space objects are all part of the solar system.

Materials

picture cards and photographs of the Sun and solar system
two index cards
straight pin
globe
flashlight
gamecards
What's Inside the Sun (Kosek 1999) or any book about the Sun

Activities

Whole Group (ELD levels 1, 2, and 3)—Introduction/background/motivation

Lesson 5

Grades 3–5

- Ask students to tell you about the Sun. Ask questions: What is the Sun? How does it affect us? Do you think the Sun is good for people? Do you think the Sun helps the planet Earth? What would happen if we had no Sun?
- Discuss how the Sun provides us with energy in the form of light. Ask students to give specific examples of how it benefits them.
- Write their responses on the board or chart paper.
- Discuss the concept of daylight and night in different areas of the world.
- Using the globe and a flashlight, demonstrate how the Sun affects day and night by shining the flashlight on one part of the globe. Remind students that it is daylight where the Sun is shining and nighttime in areas where the Sun is not shining.
- Spin the globe slowly and ask students to predict what will happen next.
- Discuss the fact that the Sun shines all the time but we cannot always see it because Earth is spinning and the Sun shines on different parts of the world at different times.
- Discuss how this affects their climates and lifestyles.
- Read *What's Inside the Sun*.

Level 1—Beginning

Target vocabulary: solar system, spiral, temperature, galaxy, energy, light

Guided Instruction

- Review the solar system (Sun, planets) and show how the Sun is the central and largest body of the solar system.
- Write the word *Sun* and draw a big circle around it. Write and introduce the vocabulary words and describe how each relates to and describes the Sun.
- Introduce new vocabulary by using visuals, writing the words, and demonstrating them. For example: Demonstrate what a spiral is (draw a spiral on the board, and make a hand motion of spiraling). Ask students to make a spiral motion with their fingers or hand.
- Introduce the word *temperature* by discussing today's temperature (talk about hot and cold, ask students if they think the Sun is hot or cold).

Independent Activity

- Have students draw the Sun. Inside the Sun, have students write words or draw pictures about what the Sun means to them.
- Have students discuss how the Sun affects them. Write down their responses on sentence strips and have students copy them and read them on their own.
- Have students use the word cards and match the definitions or pictures with the words.

Levels 2 and 3—*Early Intermediate and Intermediate*
Target vocabulary: galaxy, spiral, hydrogen, helium, pressure, photosphere

Lesson

Grades 3-5

Guided Instruction
Preview and preteach target vocabulary words.

- Write the word on the board.
- Write the word in the sentence as it was used in the book.
- Ask students to predict what they think each word means or represents.
- Write their responses next to each word in a cluster or graphic organizer.
- Write and read aloud the definition of the word.
- Review the responses on the graphic organizer and cross out or erase responses that do not apply or define the vocabulary word.

Example 1: Hydrogen

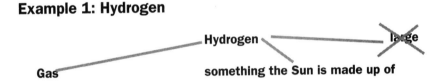

The Sun is made up of many different gases. About three-quarters of the Sun is made up of hydrogen gas.
Definition: A gas that burns easily. It is the lightest gas.
(After reading and writing the definition, review the responses and cross out or erase the incorrect ones. Example: *large* does not describe hydrogen.)

Comprehension Strategy
Write the title of the first chapter from the book (*What's Inside the Sun*, p. 5–7) on the board or chart paper. Ask students to predict what they think the chapter will be about. Write their responses on the board.

Read the chapter (a total of four paragraphs) to the class.

Ask students again what the main ideas were.

Refer to their previous responses, and cross out or erase any response that was not correct. Circle the correct predictions.

Discuss the main ideas and come up with one or two main concepts from the chapter.

Lesson
5
Grades 3–5

Independent Activity

Group students into four separate groups and assign one chapter from the book to each group.

Give each group chart paper or poster paper. Have them write the title of their chapter on the top.

Repeat the above comprehension strategy activity (ask them to predict what the chapter will be about, write down responses, read the chapter, discuss and review the chapter, and then come up with one or two main concepts from the chapter).

Level 3—Intermediate
Independent Activity

- Ask students to summarize the main concepts from each chapter. Have students write a paragraph for each of the chapters.
- Pair students, and have them read and discuss their summaries with each other.

Whole Group—Closure: ELD Levels 1, 2, 3

- Review facts about the Sun.
- Discuss how the position of the Sun affects life on other planets. On the board, write the names of the planets and the distance they are from the Sun (Pluto—3,688 million miles; Neptune—2,794; Uranus—1,784; Saturn—887; Jupiter—483; Mars—142; Earth—93; Venus—67; and Mercury—36).
- Use yarn to represent the miles. Cut the yarn the following lengths to represent the planets' lengths. (Pluto—30 feet; Neptune—23 feet; Uranus—15 feet; Saturn —7 feet; Jupiter—4 feet; Mars—14 inches; Earth—9 inches; Venus —7 inches; and Mercury—3 inches.
- Have a student represent each planet by extending that planet's yarn as far as it will reach from a spot representing the Sun.
- Have students form nine groups, one for each planet. Have each group work together to write, draw, or discuss the planet's environment and distance from the Sun. Have them discuss how the absence or presence of sunlight affects the planet.

Summary

This lesson is effective because it integrates English language proficiency levels with science content standards. It provides teachers with guidance for differentiating instruction for English learners based on the commonly used set of descriptors of English proficiency levels: beginner, early intermediate, and intermediate.[3] (See notes at end of this chapter.) This is especially helpful for teachers who are well versed in science but less so in the area of English language acquisition.

Wind Cycle

Monica Colucci, Marisela Ceballos, Miosotys Smith, Everglades K–8 Center, Miami-Dade County Public Schools, Florida
Okhee Lee, University of Miami, Coral Gables, Florida

Lesson 6

Grades 3–5

Science and Language Standards

- National Science Education Standards
  Science as inquiry
  Physical science
- Florida Sunshine State Standards in Science (Florida Department of Education 1996b)
  Strand B: Energy
  Strand D: Processes that shape the Earth
  Strand H: The nature of science
- Florida Sunshine State Standards in Language Arts (Florida Department of Education 1996a)
  Many of the standards in reading, writing, speaking, listening, and viewing apply.

Background

The lesson on wind cycle simulation is from the "Weather" unit developed by the "Science for All" research project (National Science Foundation, Grant No. REC-0089231). The research focuses on promoting science learning and English language and literacy development with elementary students from diverse languages and cultures. The research team (Okhee Lee, principal investigator) collaborated with participating teachers to develop a series of science units. These units, along with teacher professional development, demonstrated effectiveness in improving students' science and literacy achievement, as well as closing gaps among demographic groups.

In this unit, students looked at differences in air temperature. Another lesson, "How does wind form?," shows how different temperatures cause the air to move and how this movement of warm and cool air causes wind.

Vocabulary

influence, movement, simulate, wind simulation, rise, fall, upward, downward, steam, smoke

Lesson 6

Grades 3–5

Materials

a transparent box
a pair of scissors
masking tape
a smoke source (incense)
steam source (a cup of hot water)

cardboard
two plastic tubes
a hot plate
matches

Introduction

A gust of wind blew Delphine's papers all around the schoolyard. Her friend, Pierre, ran to help her gather up the papers. Then, the branches on the trees began to move. Pierre said, "I wonder how wind is formed." "I don't know, but we can ask the teacher," replied Delphine. Do you know the answer to Pierre's question?

	Wind Simulation
1. Questioning **?**	**State the problem** How is wind formed?
2. Planning 	**Make a plan** We will find out how wind is formed by doing a simulation activity with our teacher. We will use a box with chimneys, a smoke source, and steaming hot water to investigate if temperature differences cause air movement.
3. Implementing	**Gather the materials** 1. a transparent box 2. a pair of scissors 3. masking tape 4. a smoke source (incense) 5. a steam source (a cup of hot water) 6. cardboard 7. two plastic tubes 8. a hot plate 9. matches
	Follow the procedures 1. Place the clear box on the table. 2. Fit the cardboard on top of the box. 3. Cut out two circles on the cardboard. 4. Make two chimneys by fitting a plastic tube in each circle. 5. Seal any space around the two plastic tubes with masking tape.

<table>
<tr>
<td></td>
<td>

6. The teacher will light the smoke source and place it under one chimney inside the box.
Observe the movement of the smoke:
- Inside the box
- Outside the box

7. The teacher will place steaming hot water under the other chimney inside the box.
Observe the movement of the smoke and steam:
- Inside the box
- Outside the box

</td>
</tr>
<tr>
<td></td>
<td>

Observe and record the results
In the boxes below, draw a picture of what you observed in this activity. Show what happened: (a) when only the smoke was in the box, and (b) when the smoke and steam were together in the box.

only smoke in box   **smoke and steam in box**

</td>
</tr>
<tr>
<td>

4. Reporting

</td>
<td>

Share the findings
1. In your group, talk about the steps in the activity.
2. Write complete sentences to explain the events shown in the pictures. Use key words, such as *upward, downward, left,* and *right.*

</td>
</tr>
</table>

Summary

The teacher builds on students' prior knowledge because this lesson directly follows another lesson on differences in temperature. The teacher builds vocabulary throughout the lesson; for example, the teacher points to the smoke coming out of the chimney as she scaffolds by asking: "How is smoke different from steam?" or "What is the smoke doing?" The listening-while-observing, group discussion, and reporting strands of the lesson allow ELLs opportunities to practice and experiment with the English language in a context-rich and academically rigorous activity.

Lesson

7

Grade 4

Changes in Weather

Maria Aida Alanis, Austin Public Schools, Austin, Texas

Context

This 45-minute science lesson was prepared for a fourth-grade ESL classroom. The students were native Spanish speakers at an intermediate ESL level.

Standards and Objectives

The ESL and science standards are from Texas Essential Knowledge and Skills (TEKS) (Texas Education Agency 1998).

ESL Standard: Goal 2: Use English to achieve academically.

Standard 1—Use English to interact in the classroom.

Standard 2—Use English to obtain, process, construct, and provide subject matter information in spoken and written form.

Science Standard: The student knows that change can create recognizable patterns.

Identify patterns of change such as in weather, metamorphosis, and objects in the sky.

Background

In this fourth-grade ESL self-contained classroom, students are learning about weather and how weather changes. The following is the first of five daily lessons. The classroom teacher supports social and academic English through the activities and lesson design. The learning expectation is that of high-cognitive demand with high support for the learner.

The teacher designs her lessons by using a modified 5E lesson format. She includes scaffolding techniques such as modeling, bridging, contextualization, schema-building, text representation, and metacognitive development.

Activities

Engagement

The teacher uses modeling, collaborative groupings, graphic organizers, and support text selections. The classroom teacher engages the students in sessions in which children talk about the weather. The discussion includes different types of weather and weather in different regions.

Modeling

The teacher models correct English as she or he guides the students in describing the pictures in the text. The text is *Weather* (Eyewitness Books, No. 28) by Brian Cosgrove. This book uses photos and illustrations of weather conditions.

Lesson

7

Grade 4

The teacher activates prior knowledge by guiding the students through an activity using a K-W-L chart.

The teacher introduces new content vocabulary by using a graphic organizer chart with two headings: "What I think this word means" and "Confirmations of the definition."

Bridging

Students write the new vocabulary on small index cards. They draw pictures of the terms that they know.

Optional activity: Students can work in think-pair-share groups for a vocabulary development activity. They think about their responses and then share and discuss their responses with their partners.

Use schema building words and semantic webs on this weather unit.

Explore, Explain, Elaborate

- *Explore:* Have students work in small groups and use two thermometers, white construction paper, and black construction paper to find out if the color of an object affects the amount of light energy that it absorbs.
- *Explain* (metacognitive development): Have students record in their science notebook their data and construct a reasonable explanation. This is an opportunity to make their learning process explicit.
- Transition to weather conditions and have children sing the song, "Weather Forecast" found in *Avenues* (Short et al. 2003).
- Working in groups, students should record different temperatures of major cities from the newspaper.
- *Elaborate* (text representation): Have the students write a weather news summary. Have students use vocabulary such as *storm, tornado, hurricane, flood,* and *drought.*

Assessment

Informal: K-W-L, student notebook entry predictions.
Formal: Performance-based, on oral and written answers to questions on changes and variation in weather.
Follow-up question: How do different surfaces on Earth, such as water, forests, and pavement, affect air temperature?

Lesson

7
Grade 4

Summary

This lesson is a good example of a hands-on activity guided by the teacher to focus on conceptual development but which also incorporates language usage with think-pair-share activities, K-W-L charts, and student writing.

Lesson

8
Grade 5

Mixtures and Solutions

Kevin Reardon, McKinley Elementary School, El Centro, California

Topic

Mixtures and solutions

Context

This lesson was taught to a group of 20 students in grade 5, 19 of whom were Hispanic. Twelve were English learners at various stages of English proficiency.

Standards and Objectives

Science Content

This lesson partially meets California Science Standards in Physical Science for Grade 5 (California Department of Education 2003, p. 14). Specifically it meets Standard 1, as follows.

Elements and their combinations account for all the varied types of matter in the world. As a basis for understanding this concept:

- Students know that during chemical reactions, the atoms in the reactions rearrange to form products with different properties.
- Students know that differences in chemical and physical properties of substances are used to separate mixtures and identify compounds.

Investigation and Experimentation

Identify a single independent variable in a scientific investigation and explain how this variable can be used to collect information to answer a question about the results of the experiment.

English Language Development
- Ask and answer instructional questions with some supporting elements.
- Write an increasing number of words and simple sentences appropriate for language arts and other content areas (science).

Lesson
8
Grade 5

Background

This lesson is part of a unit on substances and their characteristic properties. By the end of the unit students are expected to know that a mixture of substances often can be separated into its original substances using one or more materials and that substances react chemically in characteristic ways with other substances to form new ones. This lesson from the unit on mixtures and solutions leads students to understand that concentration is the amount of material dissolved in a measure of liquid and that volume is a three-dimensional space occupied by liquid.

Materials

water, soft-drink mix
2 large containers, one green, one blue (or use 2 different colors)
paper cups (2 per student)
student science notebooks
beaker

Activities

Engagement

The teacher proposes to students that the focus question for this lesson will be "What is concentration?," then asks students what they know about concentration. (Expect answers such as "It has more flavor.") The students predict what they might be exploring in science this day. The teacher asks students to record the focus question in their science notebooks.

Exploration

- Student helpers distribute two cups to each member of the class.
- The teacher shows the class materials to be used and asks for names of each (i.e., they'll use a beaker to measure the water to be used in the mixture).
- The teacher measures 1,000 ml of water into each of the two large containers, fills 3 spoonfuls (25 ml each) of concentrated soft-drink mix, and adds it to the green container.
- The teacher asks students to look back to their written predictions and see if they match what they find when the teacher pours some liquid from each

Lesson
8
Grade 5

of the containers into two different clear cups (based on color, for example, students should be able to see some difference in the two liquids).

- The students observe the mixture in the green container and discuss what they observe from that sample.
- The students observe the mixture in the blue container and discuss what they observe from that sample.
- The teacher records collective observations from group on the board and follows up as necessary with questions such as "How did you know that?"
- Students copy observations from the board.
- Repeat the experiment with a different amount of water (500 ml).
- Students predict what they'll find, and the teacher asks for volunteers to share their predictions.
- Students record their findings once again.

Explanation

- The teacher asks for discussion about differences between the first and second experiments.
- The teacher asks students to verbalize their understanding of the findings of the experiments relative to concentration (the more mix added, the more concentrated the mixture becomes). This is followed by students writing their claims in their science notebooks.
- Students write their conclusions from their findings.

Elaboration

- The teacher refers students to the data from the two experiments and discusses the level of concentration in each, then explains the concept of concentration.
- Students must go to the next page in their science notebooks and record what they learned about concentration during the lesson.

Evaluation

This assessment is informal with a review of the student responses and questions during the lesson and then a review of student writing in science notebooks.

Summary

This lesson represents the infusion of language functions[*] (see notes at end of chapter) into science instruction as well as experimenting with different mixtures that are tested, analyzed, and reported on orally and in writing by students.

Volume and Density

Lesson

9

Grade 6

Maria Aida Alanis, Austin Public Schools, Austin, Texas

Context

This 45-minute science lesson was prepared for sixth-grade ESL students. The learners were native Spanish speakers whose ELL category level was intermediate.

Standards and Objectives

The ESL and science standards/objective are from Texas Essential Knowledge and Skills (TEKS) (Texas Education Agency 1998).

ESL Standards

Goal 2: Use English to achieve academically.
- Standard 1—Use English to interact in the classroom.
- Standard 2—Use English to obtain, process, construct, and provide subject matter information in spoken and written form.

Science Standards

- The student uses scientific inquiry methods during field and laboratory investigations.
- The student uses critical thinking and scientific problem solving to make informed decisions.
- The student knows that substances have physical and chemical properties.

Content Objectives

- Objective 1: The student will demonstrate an understanding of the nature of science.
- Objective 3: The student will demonstrate an understanding of the physical science.

Language Objectives

- Form and revise questions for investigations, including questions arising from readings, assignments, and units of study (6–8).
- Generate ideas and plans for writing by using prewriting strategies such as brainstorming, graphic organizers, notes, and logs (4–8).

Lesson
9
Grade 6

Background

Sixth-grade students receive content instruction from a content-specific instructor (science teacher). In this science ESL class, students learn about physical properties and changes. They previously learned about physical properties such as color, shape, length, and mass and will now view a lab activity that demonstrates volume and density as properties of matter.

Activities

Engage

The teacher uses modeling, collaborative groupings, realia, and support text. The teacher engages the students in sessions in which students talk about what they know about sorting objects into groups. The teacher uses articles that are familiar to the students and from their cultures or are of special interest. The activity supports using real articles to preview the content. The content topic entails learning how matter can be described according to properties and how changes in these properties occur.

Modeling

The teacher activates prior knowledge by asking if the students have separated items in an activity such as sorting laundry by color. The teacher lists students' ideas on a chart.

The teacher models correct English as she or he guides the students in describing the items or articles to be sorted into categories or groups.

The teacher introduces the lesson vocabulary: *physical property, matter, physical change, density,* and *states of matter.* The teacher lists the words on a content-area word wall where students can see and read them as they study the lesson. The words serve as a reference for students as they write or interact verbally.

Bridging

Students write the new vocabulary on small index cards. They draw pictures of the terms that they know. Because ELLs benefit from using visual scaffolding, the language is made more understandable by the use of drawings or pictures. The student is able to connect the spoken English word and the visual image.

Optional Activity

The children can work on a vocabulary development activity in think-pair-share groups.

Using a science notebook, students create concept maps for each of the terms introduced in this lesson.

Exploration

Before the activity starts, the teacher will post the following questions on a chart.

1. What is the volume of the clay?
2. What is the density of the clay?
3. How would the density change if you used a larger piece of clay?

The teacher explains that the lab will demonstrate volume and density as properties of matter. She identifies the materials that will be used: water, 100 ml graduated cylinder, 10 g clay, string.

Working in small groups, the students pour water into the graduated cylinder to the 50 ml mark. They shape the clay to fit into the graduated cylinder.

The students cut a piece of string and attach it to the clay. They use the string to lower the clay into the graduated cylinder. The teacher encourages safety by telling the students not to splash out any water.

Explanation

Metacognitive development: Students record in their science notebook their predictions of the lab activity. They can work with a partner and respond to the question: "What is the volume of the clay?" Working individually, students answer the three questions proposed in the "Exploration" step.

Elaboration

Metacognitive development: To encourage students to use their skills and vocabulary to discuss their thinking process about physical properties, the teacher has them examine items with the same volume and different mass—such as two balls that are the same size but not the same mass. The teacher encourages the students to record their observations in their science notebook or learning logs. The students can also hold a discussion with other class members and use think-alouds in which they explore and clarify concepts through questioning, hypothesizing, making deductions, and responding to others' ideas (Gibbons 2002) as they examine items with the same volume and different mass.

Evaluation

Informal: Student notebook entry predictions and illustrations.
Formal: Performance-based, on oral and written answers to questions.

Lesson
9
Grade 6

Summary

This lesson incorporates numerous scaffolding techniques—such as modeling, bridging, contextualization, schema-building, text representation, and metacognitive development—that are important for language development.

Lesson
10
Grade 7

Building Cells

Denise Minnick, Waggoner School, Imperial, California

Topic

Building cells

Context

This lesson was taught three times to different groups of seventh-grade students. Each group had from 30 to 33 students, with an average of 70% Hispanic and an average of 56% English learners at various stages of English proficiency.

Standards and Objectives

Science Content

This lesson meets the Science Content Standards for California Public Schools, Grade 7, Cell Biology 1 a,b,c,d (California Department of Education 2003, p. 22).

All living organisms are composed of cells, from just one to many trillions, whose details usually are visible only through a microscope. As a basis for understanding this concept:

- Students know cells function similarly in all living organisms.
- Students know the characteristics that distinguish plant cells from animal cells, including chloroplasts and cell walls.
- Students know the nucleus is the repository for genetic information in plant and animal cells.
- Students know that mitochondria liberate energy for the work that cells do and that chloroplasts capture sunlight energy for photosynthesis.

Lesson
10
Grade 7

Investigation and Experimentation

- Scientific progress is made by asking meaningful questions and conducting careful investigations. As a basis for understanding this concept and addressing the content in the other three strands of science (physical, Earth, and life), students should develop their own questions and perform investigations. Students will:
    - Select and use appropriate tools and technology (including calculators, computers, balances, spring scales, microscopes, and binoculars) to perform tests, collect data, and display data.
    - Construct scale models, maps, and appropriately labeled diagrams to communicate scientific knowledge (e.g., motion of Earth's plates and cell structure).

English Language Development

- Develop scientific vocabulary as well as content vocabulary related to the unit on cells.
- Write an increasing number of words and simple sentences appropriate for language arts and other content areas (science).

Background

This lesson intends to have students understand that cells are three-dimensional rather than the two-dimensional representations normally found in diagrams and under microscopes. Students should be able to note differences and similarities between plant and animal cells. This lesson requires a minimum of three class periods of about 45 minutes each.

For this lesson to be successful for English learners, there must be some advance preparation. Address the vocabulary relating to a cell and basic concepts of both plant and animal cells. This will alleviate confusion about the fruits representing elements of a cell.

Vocabulary should be recorded on a word wall, clearly presented for students to view as needed. Students should have the words listed creatively in their science notebooks, either as handouts from the teacher or as a drawing done by the students themselves. These words include *amyloplast, ATP (adenosine triphosphate), cell membrane, cell wall, centrosome, chlorophyll, chloroplast, christae, cytoplasm, Golgi body, granum, lysosome, mitochondrion, nuclear membrane, nucleolus, nucleus, photosynthesis, ribosome, rough endoplasmic reticulum, smooth endoplasmic reticulum, stroma, thylakoid disk,* and *vacuole.*

Students will participate in jigsaw reading[5] (see notes at end of chapter), paired reading, and modeled reading for this unit.

Lesson

10

Grade 7

Materials

Knox gelatin

canned fruit

square containers

plastic knives

mixing spoons

lemon Jello

boiling water

baggies and ties

drawing paper

pen and paper

Additional ideas for graphics, activities, and software to use for this lesson are available at *www.enchantedlearning.com* and *http://youth.net*.

Activities

Engagement

Students create two types of cell models (plant and animal cells). The students make models using baggies, square containers, Jello, and canned fruit to represent parts of the cell. After the students design the cells, they let the cells cool overnight for experimentation the next day.

Exploration

Students look at their Jello cells and observe the similarities and differences in their animal and plant cells. Students cut their cells to understand better their depth and the cross-sectional elements by exploring the three-dimensional elements.

Students further their understanding of plant and animal cells by labeling plant and animal cell diagrams using information on the web, such as that at *www.enchantedlearning.com*.

Explanation

Students write about the properties of plant and animal cells in their science notebooks, reflecting the idea that cells are three-dimensional. The teacher makes sure they understand that viewing cells in a book, picture, or even under a microscope shows only a two-dimensional view and that cells are three-dimensional.

Students make inferences about cells. They state what they think is happening in the cells or what they think will happen to cells. Students might further explore and predict what illness they think a sick cell might have and consider whether it might be lacking in one element or have too much of another.

Elaboration

Working in groups, the students create a three-dimensional model of an element of the cell such as the nucleus or mitochondria and explain the importance of the element to the cell as a whole. They write a report and present their conclusions to the class, using the following outline for the report: observing, questioning, hypothesizing, predicting, planning and implementing investigations, interpreting findings and drawing conclusions, and communicating (presenting).

Evaluation

For a preassessment, the teacher gives students a diagram of a plant cell and an animal cell. The students should determine which cell is plant and which is animal. The diagram provides all the elements in the cell, but the students will need to determine which element belongs where. The posttest should include the same diagrams without any of the elements provided and a question about the three-dimensional nature of cells.

Summary

This lesson combines internet resources with classroom experimentation to get students to explore various aspects of cells. Students use both oral and writing skills to report their findings. Working in groups provides peer support that makes success more likely.

Lesson
10

Grade 7

Lesson
11

Grade 7

Human Body Systems

Hafedh Azaiez, Fonville Middle School, Houston, Texas

Topic
Understanding human body systems

Context
This lesson was prepared for a seventh-grade science class. Among the 26 students, four were newcomers who had arrived within the year, eight were intermediate to advanced ELLs who had arrived within the last three years, and the rest were proficient English speakers. All were Hispanic and spoke only Spanish at home.

Lesson
11
Grade 7

Standards and Objectives

Science Objectives

Objective 9, Grade 7, TEKS (Texas Essential Knowledge and Skills) (Texas Education Agency 1998) of the Texas curriculum:

- Identify the systems of the human body.
- Describe the functions of the body system.

Language Objectives

- Students will be able to learn some key words such as *digestive, muscular,* and *immune system.*
- Students will be able to analyze the relationship between different body systems.

Background

Provide an introduction to the human body system before beginning this lesson. Introduce students to the different body systems through videos and hands-on activities such as tracing their circulatory systems and making a body systems model including digestive, respiratory, circulatory, urinary, and reproductive systems. Students should have had some experience in using websites to get information.

Materials

activity sheets   computer with internet access   pencil

Activities

Engagement

Students are engaged by working with computers. Have them practice finding specific websites and identifying what they are. Give students five minutes to visit their favorite website and describe what it is.

Exploration

Divide students into groups of two. If possible, pair each ELL student with an English-proficient student. Have the pairs brainstorm different body systems and their functions.

Explanation

After distributing the activity sheet (sample follows), go over the steps of the activity by showing the students how to get to the website and how to answer the questions by doing at least two examples. If possible, use an overhead projector so that the students can easily see the directions and steps to follow. Have the students use their own words to describe on the activity worksheet how each body system functions.

Sample Activity Sheet

1. Go to the following website: *www.ama-assn.org/ama/pub/category/7140. html.*

    Using the website, list all the body systems, their functions, and at least one organ of each system on the following table:

Body system	Function	Organ

2. Go to the following website: *http://users.tpg.com.au/users/amcgann/body/.* To answer the following questions, you must watch and read all the information provided on this website.

- How many muscles make up our muscular system?
- What controls our voluntary muscles?
- What is the function of our skeletal system?
- How many bones make up our skeletal system?

3. Go to this website: *www.medtropolis.com/VBody.asp.* After checking the brain and the heart sections, list four facts that you learned. You can also read the material in Spanish.

Elaboration

Students can make flash cards with body systems and functions. Flash cards can include pictures that can help ELL students understand and memorize a concept.

Evaluation

After finishing their flash cards, students can go through them three to four times, depending on how much time is left in the class.

Summary

This lesson helps English language learners because it relates to them, has many practical applications, and includes peer tutoring. Using computers provides students with many resources beyond the classroom.

Lesson 12
Grade 8

Understanding Water Systems
Pollution and Conservation

Maureen Sims, Jane Hill, Catherine Little, Jane Sims, Toronto School Board, Toronto, Canada

Topic

Understanding water systems: pollution and conservation

Context

This lesson was prepared for an eighth-grade class of 33 students. Eight members of the class were native speakers of English; five others had been in Canada for several years. Twenty-two were English language learners at varying stages of proficiency. They spoke Spanish, Tagalog, Korean, and Arabic. Several had arrived within the last month, others over the last two or three years.

Standards and Objectives

- Examine how humans use resources from the Earth's different water systems and identify the factors involved in managing these resources for sustainability. *The Ontario Curriculum: Science and Technology. Earth and Space Systems: Grade 8—Water Systems.* 1998. Ministry of Education and Training.

- Participate in social and academic discussions using short phrases and short sentences. (Stage 2) *The Ontario Curriculum: English as a Second Language and English Literacy Development. A Resource Guide.* 2001. Ministry of Education and Training.

- Make notes in some detail on familiar topics. (Stage 3) *The Ontario Curriculum: English as a Second Language and English Literacy Development. A Resource Guide.* 2001.

Background

This is an early lesson in a unit on water systems. The students have just completed an interview with an adult using questions such as these: "When you were young, where did your water supply come from? How did the water come to your house? How did your family make sure your drinking water was safe? How did your family conserve water? How did weather affect the water supply? How was wastewater dealt with? What responsibility did the local government have in regulating water supplies?"

Students have been instructed to bring notes in either English or their first language.

Materials

chart paper

paper for making posters

magic markers

colored pencils, pastels

Activities

Engagement

Check the notes students have taken on their interviews and then ask them to do a prediction exercise. Distribute a sheet with five questions, such as these:

How many of us do you think interviewed:

	Predictions	Number of Interviews
a person over 60 years old		
someone who got water from a well		
someone who often swam in a river		
someone who sometimes boiled water		
someone who got water from a lake		

First, have students predict how frequently they think each of these interviews would occur and then ask students to raise their hands so the class can compare their estimates with the actual figures.

Exploration

Divide students into groups of four. Have them first talk about interesting facts they discovered during their interviews. Listen carefully and note examples of pollution and conservation.

Explanation

Use the examples you have noted to introduce the terms *pollution* and *conservation.* Two topics to explore are the effects of too much rain and of the expansion of deserts. In this group, students from Egypt were familiar with the gradual expansion of the Sahara desert and the Filipino students with how flooding contaminates water systems.

Distribute chart paper and magic markers. Ask each group to use the information from their interviews to fill out a chart with examples to support each of these statements:

Lesson
12
Grade 8

- Water can be conserved in a number of ways.
- Water pollution has a number of causes.

Elaboration

Have students work in pairs to make posters about water sustainability using the ideas they have generated. Each poster must have a title, an important fact, and an illustration.

Evaluation

As students finish their posters, select two or three students to tape the completed work in the hallway, grouping similar themes together.

Finish the class with a gallery walk during which students fill out a chart with questions like these:

- How many posters show the same strategies for conserving water and for curbing pollution as your group identified?
- List ideas that are different from yours.
- In what way do you think you waste water? How can you change that?

For homework, have students use the charts to write short paragraphs using the points developed on their charts as examples to support the statements used as topic sentences.

In cooperation with their ESL teachers, some of the students will use the information gathered in their interviews for a writing exercise. A useful model could be a UNICEF publication, "A Life Like Mine: How Children Live Around the World." 2002. Dorling Kindersley Publishing, London.

Summary

This lesson is an example of how English learners can use knowledge about their community to gain a broader perspective in an integrated classroom.

Relationships in Systems of Transport

Circulatory and Digestive Systems

Ursula Sexton, WestEd, Oakland, California

Topic

This series of lessons emphasizes relationships in systems of transport, with a focus on the human circulatory and digestive systems.

The circulatory system transports materials throughout the body and interacts with other systems to maintain balance (homeostasis). When the circulatory system is out of whack, the whole body suffers.

Context

English proficiency levels: Students who participated in portions of this series had been through an intensive English language program to reach a placement level of a high L2 (high early intermediate) and L3 (intermediate) according to the California English Language Ddevelopment Test *(www.cde. ca.gov/ta/tg/el/documents/formdblueprint.pdf)*.

Two-thirds of the class were Hispanic, and one-third were of Southeast Asian origin. Most were sophomores and juniors, with two freshmen and two seniors in the group. The lesson series is appropriate for students close to exiting L2 (early intermediate), L3 (intermediate), and L4 (early advanced) levels of English development proficiency.

Science Standards and Objectives

California State Science Content Standards (California Department of Education 2003)

Physiology

As a result of the coordinated structures and functions of organ systems, the internal environment of the human body remains relatively stable (homeostatic) despite changes in the outside environment. As a basis for understanding this concept:

a Students know how the complementary activity of major body systems provides cells with oxygen and nutrients and removes toxic waste products such as carbon dioxide.

f Students know the individual functions and sites of secretion of digestive enzymes (amylases, proteases, nucleases, lipases), stomach acid, and bile salts.

Lesson Sequence **13**

Grades 9–12

Investigation and experimentation strand:

a   Select and use appropriate tools and technology (such as computer-linked probes, spreadsheets, and graphing calculators) to perform tests, collect data, analyze relationships, and display data.

c   Identify possible reasons for inconsistent results, such as sources of error or uncontrolled conditions.

d   Formulate explanations by using logic and evidence.

g   Recognize the usefulness and limitations of models and theories as scientific representations of reality.

National Science Education Standards Standard C: Matter, energy, and organization in living systems:

The complexity and organization of organisms accommodates the need for obtaining, transforming, transporting, releasing, and eliminating the matter and energy used to sustain the organism.

Big ideas:
- Systems consist of structure/function relationships;
- Systems are interrelated/integrated;
- Systems maintain homeostasis; and
- Simple systems encompass subsystems. We can identify the structure and function of systems, their processes for feedback and equilibrium, and the distinction between open and closed systems.

Language Standards and Objectives

The lesson series for high school offers opportunities for students to enhance their English language development as per the following California Standards for English Language Development (California Department of Education 2000):

Grades 9–12—Listening and Speaking Early Intermediate, Intermediate, Early Advanced

Comprehension
- Listen attentively and identify important details and concepts.
- Ask and answer questions by using phrases/simple sentences.
- Restate/execute multiple-step oral directions.
- Restate in simple sentences the main idea of oral presentations in science.

Organization/Delivery of Oral Communication
- Prepare and deliver short oral presentations, based on various sources.
- Identify key concepts and supporting details (oral presentations and familiar text).

Reading—Early Intermediate, Intermediate, Early Advanced

- Word Analysis—Word recognition, vocabulary and concept development.
- Word relationships (roots, word variations, deriving meaning from text in content area).
- Fluency and systematic vocabulary development.
- Recognize, use synonyms, antonyms, affixes, prefixes to interpret meaning.
- Read simple paragraphs/passages independently.
- Use English dictionary (use and develop a glossary).
- Use proper connections to sequence text (applied to science context).

Reading Comprehension—Early Intermediate, Intermediate, Early Advanced

Grade-level-appropriate comprehension and analysis of text and expository critique (identify, use/read fact versus opinion, cause/effect relationships in text).

- In simple sentences identify structure/format of documents (graphics, charts).
- Identify how syntax, organization, and repetitive key ideas affect clarity of text.
- Present a brief report verifying and clarifying facts using expository text.
- Read text, identify main ideas to make predictions, orally respond to factual comprehension questions, describe sequence of events.
- Listen to a brief speech and give an oral critique using simple sentences.

Writing—Early Intermediate, Intermediate, Early Advanced
Organization and Focus

- Write simple sentences to respond/exhibit factual information/connect to own experiences.
- Write increasing number of words and sentences appropriate for science.
- Write expository compositions (comparisons, problem/solutions) with main idea, and details in simple sentences.
- Collect information from various sources/take notes.
- Writing process to yield short paragraphs with supporting details (may have inconsistent use of standard grammatical forms).
- Apply recognized elements of writing techniques in own writing.
- Recognize structured ideas and arguments and support examples in persuasive writing.

Lesson Sequence **13** Grades 9–12

- Use basic strategies of notetaking, outlining, and structuring drafts.
- Investigate/research a topic and develop brief essay or report with source citations.
- Identify basic vocabulary, mechanics, sentence structure in text; revise own writing for proper punctuation, capitalization, spelling and word choice. (Each lesson includes either general objectives for language or gives reference to standards that are being covered.)

Background Knowledge Required

- Understanding the use and purpose of models, experimental use of variables and controls.
- The idea that living systems have different levels of organization—for example, cells, tissues, organs, organisms, populations, and communities.
- Basic knowledge of the structure of the cell and its functions (e.g., transport of material in and out of the cells through membranes, DNA replication for cell reproduction).

Materials

Generic system	Cards titled: transporter, medium of transport, and destination. Yarn. White boards (erasable hard surface boards) for individuals (small) and for groups (larger white boards or blank poster paper). A small set of Lego's to build a model of a house. Water, pails, and tubing of different diameters.	
Digestive system torso 3-D model handouts with blank illustration of the digestive system, and:	Upper GI (Mouth/ Esophagus):	Salt-free crackers, iodine, paper towels, wax paper or aluminum foil sheet, long inflated balloon, toothpicks.
	Stomach and helping glands:	Balances, petri dishes, bologna, crackers, spinach or lettuce and cheese, vinegar. Small freezer zipper/Ziploc bags, measuring cylinder/ cup, Vaseline, Q-tips and food coloring, 2 funnels made with thick newspaper capped at the end with a fold, medium size tub, wide bucket.
	Lower GI: Duodenum and small intestine:	Newspaper or paper towels, 2 plastic cups, 30 ml graduated cylinder, metric ruler, 10 cm² cardstock paper, tape. Inner eggshell membrane (as complete halves as possible). Begin collecting them on first day and keep eggshells in vinegar to separate the membranes easily. Rinse with water prior to using and keep refrigerated. Water, food coloring, salt, and sugar. Clear small cups, test tubes, and rubber bands.

Circulatory system poster, virtual internet models, handouts with blank illustration of the circulatory system, and:	Tubing of different diameters, containers—pails, water pumps with one-way and two-way valves. Model made by teacher connecting the circulatory and respiratory systems. (See next description and Appendix design)
Respiratory system 3-D model, handouts with blank illustration of the respiratory system, and:	A large model of the thoracic cavity showing connection between circulatory and respiratory systems, with the following materials: • clear large plastic box from cookies or licorice candy strips • 2 plastic cups • 2 one-way pumps with attachable tubing at both ends • 2 one-liter bottles of water, one tinted with blue food coloring and the other with red food coloring. (See diagram in Appendix)

Lesson Sequence
13
Grades 9–12

Models are available in educational catalogues such as Delta and FryScientific. Illustrations and diagrams are available at websites listed under Exploration 13B.

Engagement

The teacher provides cards for the students showing an example of a transportation system with its basic components: a transporter, a medium of transport, and a destination (for example, a truck, roadway, and mall; mail truck, streets, and mailboxes; microphone, wires, and speaker; elevator shaft with cables and pulleys and building floors).

The teacher gives each student one card to work on with two other peers to make a "system."

Students should:

- identify the parts of the system—name them
- identify the function of each part—explain
- identify what is accomplished by the system that cannot be done by each part alone

To emphasize the connectivity and interdependence of systems, challenge each group to find another group's system to which they can connect. They may use string or yarn to show the connection (for example, a cargo ship—ocean—port is analagous to a truck—roadway—mall).

Exploration 13A

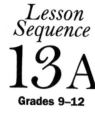

Lesson Sequence 13A

Grades 9–12

(1) To reinforce language development and to build an analogy between the interdependence, structure, and function of human body systems and building a house made of Lego's, and (2) to assess prior knowledge with this activity, using a Lego's set to demonstrate the following concepts, having students "fill in" the blanks as the demonstration is being given.

- The smallest unit in a Lego set = smallest living unit = *cell*.
- There are several different types of Lego units, depending on what they will be used for.
- When we connect several of the same kind of Lego units to make, let's say a wall or a floor, these connected sets are represented in a living body system = *tissue*.
- When we connect walls, floors, windows, etc., and make different areas/rooms for different activities, such as to do laundry, to store food, to prepare food (kitchen area), to eat dinner, etc., each of these areas or rooms is represented in a living body system = an *organ*.
- Each of these areas by themselves cannot support the entire functions of a household. If we connect areas such as the kitchen and the dining area (even if it is from refrigerator to stove to table within the same room), then we are able to have access to food storage, preparation, and consumption, representing the connection of organs to form a *system*.
- These by themselves would not have the needed electricity, space, or path to get from one to another, as well as piping or access to outside for waste disposal, such as garbage, so as to provide the comfort and necessities for an individual or family to live with some quality of life in this house. Therefore, the connection of all these areas and paths and connections to make the house functions, is analogous in a living body system to the *entire human body*.

How then are all these parts connected in ways that work systematically? How are the systems of transport explored by the students similar to or different from the analogies of the human body system?

Using a T-bar, have groups of students discuss the similarities and differences between their Lego system and the teacher's human body systems. A graphic organizer with examples follows.

Similarities between our "system" and the "house-body" system. (What is the same/ similar?)	Differences between our "system" and the "house-body" system. (What is not the same/different?)
Both systems have parts that make a whole.	*Our system* has three parts: A football play plan board, the coach, and the players in the game. *The house system has* many more parts.
Both systems show how parts connect to work together.	*Our system* shows a plan to make the players and the game work well together from beginning to end. *The house system* shows a plan for many areas, each having its own job, needs, connections, and purpose.
Both systems demonstrate how the parts function or work separately.	*Our system* demonstrates how the game board plan is a message the coach has to give to the players and then the players put it to work in the field. *The house system* shows how the kitchen is connected to the dining room, to garbage treatment, to the pipes and electricity, making it many systems and not just one.
Both systems represent how in a system there are dynamic interactions, an exchange of some sort.	*Our system* represents how a play in a game needs to have a plan and a coach to interpret and explain to the players how to make it work in the field. It is one system with a single goal. *The house system* represents many interactions, and it is many systems within a system. There are many goals. If we took only one part of the house, we would see closer similarities with our system.

Lesson Sequence

13A

Grades 9–12

Expand language use by providing repetitive patterns, sentence starters, and definitions of directives within the chart. Model on an overhead transparency how to write this graphic organizer, giving one example for each column and modeling proper English grammar.

Depending on students' language and content knowledge levels, you may need to model and complete the similarities column with them first, and have students deduce from these generic statements the specific differences of the two systems being compared, thus having them complete only the right side of the chart in their groups, or, as a whole-class exercise, using the overhead. This would be a further scaffolding step, which would avoid frustration on the part of the students.

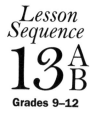

Chapter 8

**Lesson
Sequence
13 A
B**

Grades 9–12

Nonetheless, model the first row, and assess the course of action for the left column by asking for subsequent ideas from the students—if responses are given, ownership and student-centered instruction is initiated at this level. If there are no responses, provide them, allowing for the second tier of student-centered activity to take place at a more tangible level.

Depending on the nature of the types of systems the students described, clarify misconceptions accordingly. You may also bring to students' attention other types of systems not considered, such as transfer of energy through a system (i.e., nervous system resembling the wiring of a house, transfer of waves in the hearing system, or radio waves through antenna networks), identifying the transporter, medium of transport, and the destination, so as to clarify that it is not just matter (solids, liquids, and gases) that can be transferred from one location to another. The purpose is to help make connections in terms of scale, so students can begin to associate the similarities and differences that exist in the flow of matter and energy taking place at the different levels of organization of living systems and their interaction with the physical environment. What is evident at the macro level and how this is perceived is one way we can begin to understand the processes that occur at the micro level. Previous lessons on cellular functions, with transference of chemicals, and DNA replication would help to establish this conceptual connection even further.

Exploration 13B

To focus attention on transportation systems, the students will explore designing and building a model of a transportation system that can move a fluid from one place to another.

- For homework, have each student prepare ideas based on his or her background. This will validate each individual's contribution, and it will enrich the collective knowledge base of the group to which they are contributing. (Some students may relate to rural Mexican rain-gutter water collection for sustained living conditions; others may have family members who work with cars as mechanics' apprentices; while others may come from migrant communities with irrigation systems for agriculture.)

- Using the inquiry approach, ask students to work in groups to design on their white boards a way to move a fluid from one spot to another. Review models built using the criteria chart that follows. This provides a tool with a specific purpose and emphasizes the importance of data collection and record keeping to inform further steps and decisions of the exploration.

Component of design being analyzed	Model: Dump	Model: Squirt	Model: Siphon	Model: Other	Positive (+) and Negative (-) factors feedback What works /Is still needed...
Ability to move "stuff"	Yes	Yes	Yes		
Continuous flow	No	No	Yes		
Circulates (back to start)	No	No	No		

Lesson Sequence

13B

Grades 9–12

Each group presents to the class its design, and has a "forum critique" to help refine the designs, giving feedback to their peers, facilitated and moderated by the teacher with the above chart criteria. Students list the "+" and "−" in their charts and prepare for experimental design. This models to a certain extent real-world study of science and how scientists' journeys toward discoveries are peer reviewed, meet with skepticism, and require further questioning, trials, and patience.

Once students have had an opportunity to see the + and − of each system, they will be able to explore ways to use the materials provided (water, pails, tubes of different diameters, gravity) to make the fluid transport from one container to another.

To help establish the connection between their mechanical models and the biological setting, debrief the exploration by discussing the "correctness" of their models and how they might work or not work for a biological situation.

Provide new input by showing that one part of the system, such as the pail, must stay fixed; yet fluid must flow from one to another. Demonstrate with a design of a system that does so.

When you demonstrate the design, the students will recognize the need for a pump in their designs and understand the need for pump action, tubing of different diameters, and the difference between open and closed systems, as they continue to add to their + and − design annotations.

Elaboration

Assign for homework, library, or internet research, reading on how the systems designed may be similar or different from a human's circulatory system. Using a T-bar graphic organizer similar to the one in Exploration 13A, model on the overhead the specific expectations for the assignment .

Lesson Sequence 13B

Grades 9–12

The students should:

1. Show *design of own system* with *labels of each part* of the system and *how parts work.*

2. Use provided circulatory system diagram to connect with lines portions similar to system design made in class.

3. Use the T-bar on the back of the page to show similarities and differences between students' own design model of a system and the circulatory system, as shown in class.

Provide a handout as follows with a clear illustration of the circulatory system with labeled parts/functions and a blank space on the right for a design of students' own systems, as follows.

Front of handout:

Teacher-provided diagram of circulatory system (with labels and functions provided)	Student diagram of transport system designed in class (to find similarities to left-side diagram)

Back of handout:

Similarities between circulatory system and our group's designed transport system	Differences between circulatory system and our group's designed transport system

Resources for Reading Materials

Provide previewed reading materials for varying readability-language levels, and/or handouts that have been modified to identify key vocabulary with definitions embedded; function words (scientific process words) and/or diagrams that will facilitate understanding of the concepts being addressed; and highlighted text referring to explanations of specific ideas to be explored (i.e., function of each part of the circulatory system). Some resources are at *http://science.howstuffworks.com/heart3.htm, http://infozone.imcpl.org/kids_circ. htm, http://sln.fi.edu/biosci/systems/circulation.html,* and *http://vilenski.org/science/humanbody/hb_html/circ_system.html.*

These two sites are to be used in conjunction with each other: *http://vilenski.org/science/humanbody/hb_html/heart.html* and *http://vilenski. org/science/humanbody/hb_html/trachea_lungs_heart.html*.

More advanced reading sites are at *www.americanheart.org/presenter.jhtml?identifier=4567*, *www.americanheart.org/ presenter.jhtml?identifier=1557*, and (a Spanish reference site) *www.american-heart.org/presenter.jhtml?identifier=3015971*.

A printable blank diagram of the circulatory system and reading text site is at *www.lessontutor.com/jm_circulatory.html*

Elaboration and Evaluation

Post the rubric on the overhead, and direct students to work in pairs to debrief their assignment with each other. In pairs, students share their researched information with one another and exchange papers to verify if rubric steps were completed by their partner, giving feedback and comments to each other. (Monitor this to have an overview assessment of student interactions and the quality of the work/review process.) Students turn in assignment. Return it, with comments, for students to place in their binders.

Exploration 13C

To provide exploration of the relationship between the circulatory system and the digestive, excretory, respiratory, and endocrine systems, introduce each system's parts and functions, through a series of minilabs. These labs should engage all students with the process of transport of materials in these living systems, through analysis and exploration in context, using tangible tools, visualization, and recording procedures. The facilitation should include helping to make connections between the macro level of observation and experimentation, and the more abstract micro level of transfer of materials at the cellular level, as well as bridging into diseases and malfunctions of these systems. Inquiry activities related to the digestive system are outlined below.

Digestive System

Use overhead transparencies to show an illustration of a digestive system matching the handout provided to students. Model labeling of each part on the diagram and on the vocabulary glossary organizer the students will keep during the unit.

Lesson Sequence **13C**

Grades 9–12

Body Systems – Glossary – Vocabulary Organizer

Organ System part	Function / What each part does (includes drawings)		Illnesses/ problems
Mouth	Teeth	Cut, tear, grind foods into smaller pieces.	Cavities, gum disease, cancer (smoking)
	Saliva	Comes from salivary glands. It makes food easier to swallow. It has the enzyme amylase, to break starch molecules into sugar molecules. Starches + saliva/Amylase ➡ Sugars	
	Tongue		

Model use of simple sentences to describe their functions. The students copy the information after each component of the system is explored, in groups or as a class, with tangible materials as follows:

1. Distribute a sheet of wax paper or foil, 6 crackers, toothpick, and iodine to each group.

2. Ask students to leave a cracker on one-half of the wax paper/foil sheet untouched.

3. Each student breaks off a piece of his or her own cracker to test the first part of digestion, keeping the other piece for a second test to follow. Students attentively observe the functions of all the organs in their mouth as they chew and swallow.

   Prompt for words that describe what they just experienced, and write terms related to the mouth on the board or a poster for future reference (e.g., *cracked, crunched, slippery, wet, sliding, powdery, chew, swallow, pushed, pressed, moved, taste*). On the diagram–chart, model writing: *mouth* (*teeth*—tear, cut, and crunch food into smaller pieces; *tongue*—tastes food and pushes it back).

4. One student in each table takes a small piece of the center cracker and chews and spits it back onto the other half of the wax paper/foil and

adds a drop of iodine to both samples, stirring with a toothpick the chewed sample. The group observes for differences in color of the solid cracker, and a brief discussion facilitates understanding of the difference (the solid cracker turns black as an example of a chemical reaction between starch and iodine, and the chewed cracker turns purplish as the starches change into sugars by the saliva enzyme amylase.

Lesson Sequence 13C
Grades 9–12

Model on the board how to add definitions to the glossary (saliva from salivary glands—makes food easier to swallow. *Saliva*: has the enzyme *amylase*, which breaks starch molecules into sugar molecules.) Add an illustration of a chain and links broken apart.

Student who chewed and spit properly collects wax paper/foil and disposes of it, then washes hands thoroughly.

Ask: When you swallowed, do you think it was gravity making the food go down? Or could it be something else? How could we prove it? How do the astronauts swallow while in space? If no one comes up with the idea of swallowing while upside-down, have students lower their heads between their legs while chewing and swallowing the other half of their cracker. (For more dramatic results, but taking into account safety, ask if one student knows how to do a handstand and will volunteer to swallow a cracker while in that position.) The point is made, and you clarify that it is by muscle contractions in the esophagus that material is pushed into the stomach.

Writing on the glossary continues: *Esophagus*: pushes food into stomach by muscle action or *peristalsis*. (Demonstrate with a long balloon inserted into a piece of paper with a hole in the middle how the movement occurs as it is squeezed from one end to another and, when it passes a membrane that separates the lungs from the abdomen, it finally enters the stomach.) Explain that a ring of muscles at the end of the esophagus and at the opening of the stomach usually are tighter and create higher pressure than the pressure in the stomach, making the food travel in one direction. If a person has acidity, reflux, or heartburn, the ring of muscles is too relaxed, allowing acid from the stomach to come back up into the esophagus. You may ask for a show of hands of how many students have felt a burning sensation in the mouth of their stomachs. Likewise, when the body needs to get rid of food that may be contaminated or bad, it reacts by allowing a person to bring up the contents of the stomach.

5. Ask students to try to breathe and swallow saliva at the same time. They can't. Have them put their fingers on their throat while they attempt again. Have them feel just as they swallow, then again just as they breathe. Show a diagram of the pharynx and esophagus at the point

Lesson Sequence

13C

Grades 9–12

where they separate, and explain the motion they felt is a valve or trap-like door structure called the *epiglottis*—it doesn't allow food to go into the lungs. It closes when we swallow and it stays open for breathing. When we choke, something is getting trapped here, and a cough is a natural reaction. If something is wedged, we are in need of help. Ask students if they know the Heimlich maneuver.

6. Next, get four volunteers to help measure the materials that will be used to demonstrate the function of the stomach. Four more help hold and manipulate the materials so as to represent as accurate a model of the stomach as possible and to see how different foods may be processed. Each of the four bags will contain a different type of food: diced bologna (protein/fat), shredded spinach or lettuce (roughage/vegetables), crumpled crackers (starches), and shredded cheese (dairy/fat).

Using a clear and descriptive diagram of the layers of the stomach, explain how they are lined with many glands that *secrete* (produce and release) *digestive enzymes* and other materials that *help the process of digestion*. Ask the students to help fill in the blank on the statements being shown on the overhead transparency, as each item is shown:

- The Ziploc bag will represent the *stomach sac.* (Four students hold bags facing the class.)
- The Vaseline will represent the *mucus* that lubricates and protects the *stomach wall.* (Student smears with Q-tip some Vaseline on the inner lining of the four bags.)
- The vinegar will represent the *acid* that helps to break down food. (Student measures ¼ cup for each bag and pours into each.)
- The yellow food coloring will represent the *pepsin* enzyme that breaks down proteins. (A student puts a drop in each bag, letting it slide from the sides, as if coming from the lining.)
- Shaking will represent the muscle action of churning and mixing food into *chyme.*

In the meantime, two other students weigh equal amounts (2 oz.) of four different foods: shredded bologna, shredded cheese, shredded lettuce, and crumbled cracker. Ask the class why it would be important to ensure that each food weighs the same and why we need equal amounts of vinegar for this demonstration. Discuss the importance of maintaining controls and variables, so as to help visualize what might happen in the stomach with each type of food. Compare the volume differences of each type of food, despite their equal weight, and reiterate why all materials were shredded (e.g., have been processed by the mouth—though missing saliva action).

Put weighed food into each bag and seal each completely. Direct students to shake the bags an equal amount of times (x50) from side to side (modeling how to shake vigorously without dropping the bag) for the whole class to observe changes taking place. Compare each bag, and infer which foods might need more time for digestion than others. Briefly debrief the inaccuracies and limitations of the model (e.g., there are not real enzymes causing reactions of molecules to break down foods—some breakdown by acid might occur, but vinegar is not as strong an acid as hydrochloric acid produced in the stomach, nor are the amounts proportional to the production of acid in the stomach according to the amount of food entering it).

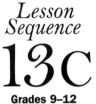

Lesson Sequence

13C

Grades 9–12

Record in the glossary the functions of the *stomach*: Stores food. Makes *chyme* by mixing foods. Produces *mucus* to protect its walls. Produces *acid* to break down foods and *pepsin* to break proteins. You may draw a set of letter P interconnected as a chain and then pieces of the letters P's broken apart, similar to starch breakdown in the mouth.)

Briefly explain the cause of peptic ulcers is the recently discovered *Helicobacter pylori.*

Use two thick newspaper funnels sealed at one end, to hold liquids representing:

* the *gallbladder bile* produced in the liver, and
* the *enzymes* produced in the *pancreas* (amylase [starch breakdown], trypsin [protein breakdown], lipase [fat breakdown].)

Demonstrate on a torso model of the digestive system the placement of these organs and their spatial relationship to the duodenum.

Pour green colored water (bile) into one funnel and pink-colored water (pancreatic enzymes) into the other. Mix all the contents of the four bags previously representing stomachs into a large Ziploc bag (one stomach) that will have a bottom corner cut (pylorus), and as its contents pour onto a basin tub, the two funnels are simultaneously released onto the basin to catch all three liquid materials into it. This represents the exit of processed materials from the stomach and the auxiliary glands' enzymes into the small intestine.

Write in the glossary the function of these two organs in relation to the process of digestion.

Liver: Makes *bile*—a chemical to help break fat apart over a larger area, so later enzymes can break them down even further (like soap on an oily cooking pot—*fat emulsification*).

Pancreas: Makes pancreatic fluid to help break down carbohydrates (starches), proteins, and fats. Makes hormones that control sugar levels in

Lesson Sequence

13$_{D}^{C}$

Grades 9–12

the blood. Makes sodium bicarbonate (like baking soda) to reduce the acid in the intestine, since enzymes can't do their job within an acid environment.

Exploration 13D

This exploration demonstrates duodenum and small intestine functions, and the breakdown and transport of materials into the blood system.

Using a torso model, briefly review with students the body parts and functions involved in digestion covered so far by having students write and draw on individual white boards a few of the parts and functions learned (formative whole-group assessment).

Ask students to look at the images on the overhead to decide which has the largest surface area (flat surface within the lines given) and to think how they could prove it. If no viable responses are given, have available a string and ask if using this tool would help and how. The string would help to compare the circumference length of A and B. If B were to be stretched into a smooth circumference, it would make a larger circle, thus having a larger diameter/radius and consequently a larger area (πr^2). Some other possible solutions: The linear length of each figure's circumference is different: A is shorter and B is longer. If a perfect square were made with these lengths, the sides of a square made from the length of A would be shorter than those made by the length from B. Therefore, the area of the square "A" (base A x length A) would be smaller than a congruent square "B." (Finding the area of a square might be easier for students for this purpose.)

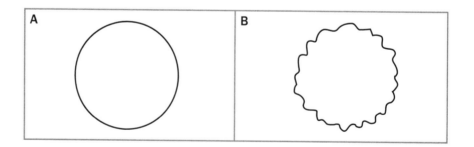

To understand how important the small intestine is, we need to understand the importance of the surface area it has. Therefore, have students compare the surface area of two cylinders made from 10 cm² cardstock paper, lined with a smooth paper towel and with a paper towel folded in zigzag.

Give each team two pieces of 10 cm² cardstock paper, two paper towels, two cups, a 30 ml cylinder, and water. Have them prepare the first cardstock by cutting a paper towel to the size of the square, laying it on it, and then rolling it into a cylinder (without overlapping) to represent the lining of the intestine, if it were a smooth wall. Then, on the next cardstock, fold in zigzag a paper towel until it fits the size of the square, cutting off the edges that hang out at the bottom. This represents the intestine with villi. Using visuals as follows, have students perform the task:

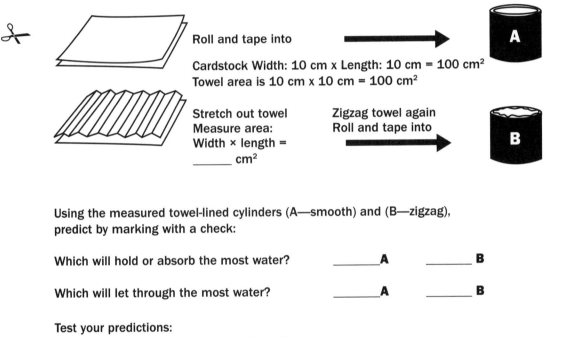

Roll and tape into → A

Cardstock Width: 10 cm x Length: 10 cm = 100 cm²
Towel area is 10 cm x 10 cm = 100 cm²

Stretch out towel
Measure area:
Width × length =
_____ cm²

Zigzag towel again
Roll and tape into → B

Using the measured towel-lined cylinders (A—smooth) and (B—zigzag), predict by marking with a check:

Which will hold or absorb the most water?   _____ **A**   _____ **B**

Which will let through the most water?   _____ **A**   _____ **B**

Test your predictions:
Slowly pour 30 ml of water into cylinder A over a cup.
Measure and record how much water is collected in the bottom cup_____ml.
Slowly pour 30 ml of water inside cylinder B over a cup.
Measure and record how much water is collected in the bottom cup _____ml.

Describe and show a diagram of a crosscut section of the villi inside the small intestine. (A "kush-ball" resembles the folds and can be used as another visual of the lining of the intestine.)

Explain there are nearly 7 meters of small intestine folded inside our abdomen absorbing and exchanging through the fingerlike projections called *villi*, and each of the villi (villus) has microscopic microvilli (about 20,000 villi and ten billion microvilli in a square inch) is in charge of making the exchange of materials across its membranes into and out of the blood stream.

Discuss in small groups and debrief: How did the folds in zigzag affect the water absorption?

*Lesson
Sequence*

13D

Grades 9–12

Why is the surface area of the small intestine important? What would happen if our small intestine was like tube A instead of tube B?

Exploration D-2

To be able to compare and explore transfer of solutions across a living membrane, two different solutions of concentrated water will be compared:

- half the class works with sugar-concentrated dark blue water, and
- the other half works with salt-concentrated dark green water.

Use the following section as overhead and as handout to model expectations and go over questions and clarifications, with the use of simple illustration icons for each step. Keep posted during the activity.

The students will build a model of material transfer across a living membrane, by using an egg membrane and testing two different concentration liquids as follows:

- **Measure 30 ml of color-concentrated water and pour into a test tube.**

- **Wrap an egg membrane at the opening of the test tube like a drum and hold it securely with a rubber band around the edges, being careful not to tear it.**

- **Measure 60 ml of clear water into a cup.**

- **Dip the tip of the test tube with the membrane touching the clear water for 1 minute.**

- Observe, draw, and record any changes observed.

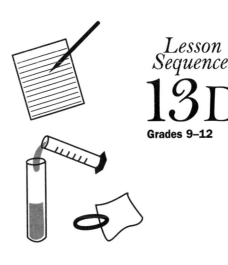

- Carefully, remove the test tube from the water cup.

- Take the membrane off and pour colored concentrated water from the test tube into the measuring cylinder. Measure the liquid content of the test tube and record. _____ ml

- Measure the liquid in the cup again and record. _____ ml

What can we conclude from comparing these two different concentrated solutions (salt and sugar) as they interact with a living membrane and are in contact with a less concentrated liquid (water)?

Which part of this model represents the intestinal *villi* and which represents the bloodstream?

Were there any changes in volume of liquids before and after connecting them with the membrane? Any observed transfer of material from one liquid to another? How did the salt and sugar concentrations' behaviors compare? Which way did the transfer occur? (Create a class graph with all the data, and allow for student responses in pairs, round robin, and whole-class response as you alternate through the questioning, providing opportunities for all to participate.)

Discuss that this model also has limitations, for we cannot see the molecular level and cellular level of the transfer from, let's say, *disaccharides* (complex sugars) to *monosaccharides* (simple sugars). Add terms to the glossary as by-products of the functions in the small intestine. However, from previous lessons and as a new application to cellular functions, students may apply the concept of transfer of substances across biological membranes in the context of the function of the villi in the small intestine. Most living membranes are selectively permeable

Lesson Sequence

13D

Grades 9–12

(allowing some materials through and not others) by the processes of osmosis, facilitated diffusion, or active transport. The transfers of these materials across cellular membranes seek homeostasis or equilibrium, moving material from a more concentrated solution (intestinal fluids) to a less concentrated one (blood stream), until both sides of the membrane are the same.

The villi bring in fresh, oxygenated blood, and the oxygen goes into the tissues of the intestinal wall. The villi send out nutrient-enriched blood, by absorbing nutrients from the liquefied food, now turned into molecular-size materials, through the membrane of the villi. Inside the villi, enzymes further break down the molecules to complete digestion and then transfer them into the blood and lymph vessels. Fatty nutrients go into the lymph vessels. Glucose and amino acids go to the blood and on to the liver.

You may use the last process of digestion to act out the functions at the cellular level. Students use labels (papers with names of proteins, disaccharides, lipids, and so forth) to create a map or diagram on the classroom floor of the semipermeable membranes and parts involved using yarn or the layout of the desks. Have students representing enzymes within the villi membrane tear or break down the disaccharides (papers being brought in the digestive fluid group of students) into monosaccharides (papers cut in half and writing glucose, fructose, galactose) or polypeptides (whole paper) into amino acids (cut papers). These can be carried or passed on with a pumping motion through to the other side where the blood vessel membrane receives the molecules and passes them into the blood to go to the liver, or the lymph vessels' membranes into the lymphatic system. This illustrates the flow of materials between two differently concentrated solutions across the membranes of the villi, completing digestion and absorption of materials for survival and growth.

Most of the absorption of materials happens in the last two segments of the small intestine (the jejunum and the ileum). As food is moved forward by waves in the intestine, what does not get absorbed passes on to the large intestine. For this, students give pieces of paper with the word *water* on them to students representing membrane cells and carry the remaining blank pieces of paper as food waste material toward the sign *colon* followed by arrows to *rectum* and out. By observing the transfer across the egg membrane, students can observe at the macro level changes taking place, and, through acting out the process, they can visualize the smaller parts that make the job possible and how the end of the digestion process involves absorption of nutrients through membranes.

As an assignment, students investigate illnesses related to the small intestine and large intestine to share with the class.

Evaluation

For homework, students illustrate and write the journey of a hamburger, telling the story of its digestive path from mouth to blood/lymph vessels with diagrams, labels, narrative, and arrows. Provide a template with a matrix of the key stages of the digestive process partially completed in class and a diagram of the human digestive system on the back. Model how to use this tool as an outline resource to write and illustrate their story. (Boldface words are those provided by the teacher; others are samples expected from students.) Tell students to refer to their glossaries and class notes for this task.

Lesson Sequence
13D
Grades 9–12

Journey of a Hamburger Through the Digestive System

Organ (label/draw)	Key Physical Stages	Key Chemical Stages
Mouth	*Chewing* *Grinding* *moistening*	**Salivary amylase** *Starch* ⟶ **maltose (simple sugar)**
Esophagus	*Moistening* *peristalsis* **(muscle push action)**	**none**
Stomach	**Moistening** *Churning* *peristalsis*	*Acid (HCl)* and *Pepsin* **Proteins** ⟶ **polypeptides**
Small Intestine	*Most food absorption happens here.* **(villi)** **Peristalsis (muscle push action)** **Fat** *emulsification* **(like soap)**	*pancreatic amylase* **Starch** ⟶ *maltose* **(simple sugar)** Proteases, e.g. typsine **Proteins** ⟶ *polypeptides* *Peptidases* **Polypeptides** ⟶ *amino acids* *lipases* **Fats** ⟶ **fatty** *acids* **and glycerol** **Maltase, lactase, sucrase** **Disacharides** *Monosacharides* **(complex sugars)** ⟶ **(single sugars)**
Large intestine	*Water* **absorption back to body** *Waste elimination* **Peristalsis (muscle push action)**	**none**

On the back, provide an image of a blank digestive system scheme and space in which to write and draw.

Lesson Sequence **13** Grades 9–12

Final Elaboration and Evaluation

In stations around the room, provide three-dimensional models of body systems, such as the heart-lung teacher-made model, a torso with the digestive and excretory system, a lung model in a jar that inflates and deflates, and a poster or 3-D diagram or model of the circulatory and the endocrine systems. Students should also have access to computers and binder records. Allow students to manipulate objects, read prepared cards or handouts, and conduct research on the computer with information about the other systems affected by the transportation of materials in the circulatory system.

Note: Key terms for functions may be *moves, transports, facilitates, to/ from.*

Show on an overhead the following expectations from this session, giving examples, using the already-studied digestive system (italics are responses expected on blank lines) for students to observe and discuss as they move around the room comparing and recording the following:

1. the function of each of the systems as they relate to the circulatory system in one sentence, such as:
    Example: Digestive system moves food to cells.
    *   *Excretory system transports waste for removal.*
    *   *Endocrine system transports hormones.*
    *   *Respiratory system facilitates gas exchange (oxygen and carbon dioxide).*
2. the parts of each system (skip digestive) and its function using provided body systems' blank illustrations; cut and paste one in each page of your binder, complete information using the computer and models, cards and handouts at each center.
3. the types of materials produced by each system such as:
    Example: Digestive system produces enzymes, bile and acid.
    *   *Excretory system produces* _____
    *   *Endocrine system produces* _____
    *   *Respiratory system produces* _____
4. how the product moves (from where to where is it being transported in each system?)
    Example: Food material is broken down starting in the mouth and it is transported through the esophagus into the stomach, breaking down more, to the duodenum (with added gland secretions from the liver and the pancreas) into the lower small intestine, through the villi into the blood and lymph vessels, continuing into the large intestine where water is taken back into the body. Finally waste is eliminated out of the body through the rectum.

Excretory system:

Endocrine system:

Respiratory system:

5. problems caused by the circulatory system not working well and how these are related to the various other systems: digestive, endocrine, respiratory, and excretory. (Make a T-chart on another page.)
6. what happens when you hold your breath and why you have an urge to exhale.

See body systems information at *www.kidinfo.com/Health/Human_Body.html* .

Summary

This series of high school activities worked for ELLs because they provided a sequenced set of supports or scaffolds to build vocabulary in context with explorations, lowering students' threshold of anxiety in the interactions and capitalizing on their prior knowledge and experiences.

Modeling the use of graphics, charts, and recording tools supports academic language development, organizational structures, and processes specific to the scientific field and contextual language development.

Glossary development in context with tangible experiences and using clear language with brief but accurate definitions and illustrations help students retain concepts, as do the review white boards and the assignments.

Teacher modeling, examples provided ahead of time, and structures and expectations given in anticipation of performance facilitate clarification of goals, opportunities to correct misconceptions, and plenty of opportunities for students to listen, speak, read, and write about science as they explore concepts.

Specific repetitive sentence structures and amplification of language use, repeating and clarifying, using synonyms and definitions in context provide meaning and clarity and better access to the key terms and concepts being studied.

Incorporating technology and preparing handouts ahead of time give support for students to learn ways to research and gather information. Gradually, as with the application and final extension activities, the teacher removes some scaffolds, and monitors and facilitates as students work in small

Lesson Sequence

13

Grades 9–12

groups and independently to finish a product or to gather information.

Use of the 5E model of instruction with its dynamic opportunities allows incorporating some new vocabulary up front, most embedded within context and some used for assignments and independent research.

The "Total Physical Response"—acting out a process of transport of materials at the cellular level—is left for last, to ensure that these more abstract concepts are clearly visualized after performing multiple activities that reinforce the concept of transport systems in the human body. The incorporation of parts in the whole and the analysis of scales comparing macro and micro events worked well in the sequence given and match the content standards for these grade levels. Students enjoyed using white boards and claimed they remembered much more by developing the glossaries and reviewing with white boards once or twice a week.

Lesson 13 is reprinted by permission of WestEd. It incorporates material from *Access to Content Standards for English Learners*, copyright 2005 by WestEd. Teachers in the K–12 Alliance Network (*www.k12alliance.net*) and the Mount Eden High School, Hayward, California, participated in its development through the Access to Content Standards for English Learners Project (ACSEL), an extension of the Map of Standards for English Learners Project, which provides related products and services at *www.wested.org/cs/we/view/rs/719 and www.wested.org/cs/we/view/serv/30.*

Appendix

Illustration of a model connection of the respiratory-circulatory systems

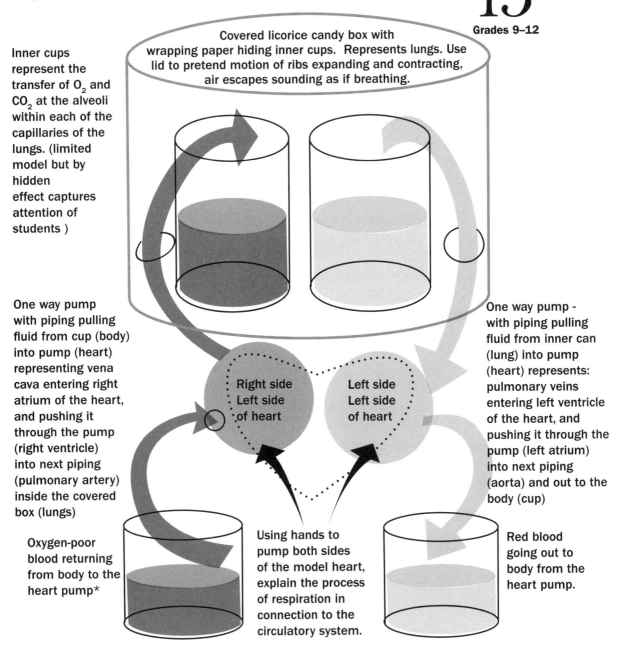

Inner cups represent the transfer of O_2 and CO_2 at the alveoli within each of the capillaries of the lungs. (limited model but by hidden effect captures attention of students)

Covered licorice candy box with wrapping paper hiding inner cups. Represents lungs. Use lid to pretend motion of ribs expanding and contracting, air escapes sounding as if breathing.

One way pump with piping pulling fluid from cup (body) into pump (heart) representing vena cava entering right atrium of the heart, and pushing it through the pump (right ventricle) into next piping (pulmonary artery) inside the covered box (lungs)

Oxygen-poor blood returning from body to the heart pump*

Right side
Left side
of heart

Left side
Left side
of heart

One way pump - with piping pulling fluid from inner can (lung) into pump (heart) represents: pulmonary veins entering left ventricle of the heart, and pushing it through the pump (left atrium) into next piping (aorta) and out to the body (cup)

Using hands to pump both sides of the model heart, explain the process of respiration in connection to the circulatory system.

Red blood going out to body from the heart pump.

* For this simulation, pumps such as those used for measuring blood pressure manually can be used.

Effective Lesson Components

These lessons focus on the development of science skills, conceptual and process skills, and the incorporation of teaching and learning strategies that are very appropriate for English learners. Many of these strategies have been discussed in Chapter 5. Some of the components that make them especially effective for ELLs include:

Integrating science and language: The lessons integrate science and language. Lessons are based upon specific science and language standards or objectives. All of the lessons provide hands-on experiences that allow students to interact with others and demonstrate their understanding of science content through oral, written, and graphic forms of representation.

Building on students' prior knowledge: The lessons are all part of larger thematic units. Many lessons involve reviewing prior knowledge. The concepts, vocabulary, and activities build upon what has been learned and experienced previously by students, both in and out of the classroom.

Using hands-on inquiry: In the lessons, students are involved in working with concrete materials that can be manipulated and described. Students go through a series of steps that include planning, predicting outcomes based on scientific information or prior knowledge, observing, and recording findings, followed by discussing and reporting of the implications of their findings

Introducing key vocabulary: The teachers build vocabulary by using it in context, introducing key words for lessons, and providing scaffolding by modeling and asking questions.

Providing scaffolding: This is exemplified frequently in all lessons when students answer questions posed by the teacher or repeat aloud or in writing information given by the teacher. The many support systems such as modeling, bridging, contextualization, text representation, and metacognitive development are important for language development.

Developing key science concepts: The lessons allow ELLs to develop key concepts and practice and experiment with English language in context-rich and academically rigorous activities.

Building knowledge based upon essential concepts or big ideas: Students learn big ideas about real-world phenomena—such as change, relationships, cycles, and systems—with practical applications.

Using cooperative groups: Students work in groups to discuss their observations, share their explanations, and write their findings. Working in groups gives ELLs the opportunity to share their knowledge and develop social and academic language through the help or modeling of students more proficient in English.

Assessing learning: Lessons include examples of both oral and written assessment through questioning, discussions, science notebooks, and reports. Assessment strategies are flexible and varied based upon student backgrounds and proficiency levels.

Addressing the needs of different proficiency levels: The activities in these lessons are of varying degrees of difficulty. Students are provided with opportunities for guided support from the teacher and help from peers in group and pair work. All students are encouraged to participate in activities in varying degrees, irrespective of their English proficiency.

The components listed here reflect some of the most common best practices teachers use when integrating science and language learning. But novice teachers may feel overwhelmed if they believe that they need to incorporate all of these components at once into each lesson. They need to understand that they can choose only the most appropriate components—some may lend themselves better to certain lessons than others. The teachers who developed our sample lessons have demonstrated how to incorporate some aspects of these practices into their lessons, not all in every lesson.

With a careful eye to the potential integration of these strategies into lessons, eventually a teacher can integrate components that will enhance language learning into all lessons. For example, if a teacher usually teaches science lessons that integrate reading but neglects to integrate writing or speaking activities, then the instructional program will not be optimal. If, however, the instructional program has lessons that focus on reading one day, others that lend themselves well to writing the next, then there is a better chance of benefit over time for the students.

Conclusion

This chapter provides us with a glimpse into how lessons at various grade levels can be rich in science content and process and in pedagogical strategies for language learning. Attention has been given to essential elements that make science learning accessible to English learners. Whether in an elementary or secondary setting, teachers can design lessons that have as a goal the inclusion and participation of all students and that are especially sensitive to the linguistic needs of English learners.

Notes

1 K-W-L is an acronym for Know, Want to know, and Learned and is a strategy commonly used to first document what students know about a topic (determine prior knowledge), what they wish to learn about it, and subsequently, what they learned about it.

2 Think-pair-share is an instructional strategy in which the teacher asks each student to think individually about an idea, problem, or task, then to pair up with another student and share their thinking about the topic. The teacher may elect to have some students share their thoughts with the entire class.

3 Various states use these or similar designations to describe student levels of proficiency in English. These are often aligned with levels determined by assessment programs approved for use by state education agencies. In California, such designations stem from the use of the California English Language Development Test (CELDT). See *www.cde.ca.gov/ta/tg/el.*

4 Language functions refer to language domains or categories of language that children use for a variety of purposes (Christie, Enz, and Vukelich 1997). It also encompasses skills students use to perform academic tasks, such as comparing and contrasting, that require higher-order thinking. In the case of English learners, it also means the development of vocabulary enabling them to convey their knowledge resulting from activities such as comparing and contrasting .

5 Jigsaw reading (Echevarria, Vogt, and Short 2004) refers to a strategy in which students are grouped to work collaboratively and one, or two students from each group reads one section of the text and becomes the "expert(s)" for that section. Students are then regrouped so that each expert can share knowledge with students who did not read the section.

References

Bassano, S., and M. Christison. 1992. *Life science: Content and learning strategies.* Reading, MA: Addison-Wesley.

California Department of Education. 1997, 2003. English language development standards for California public schools, kindergarten through grade twelve. Sacramento, CA: Author. Available at *www.cde.ca.gov/be/st/ss.*

California Department of Education. 2003, 2000. Science content for California public schools, Kindergarten though grade twelve. Sacramento, CA: Author. Available at *www.cde.ca.gov/be/st/ss/.*

CISC Science Subcommittee. 2002. *Strategic science teaching, grades K–12, A sampler of science lessons connecting literature with California standards.* Sacramento, CA: California County Superintendent Educational Services Association

Christie, J., B. Enz, and C. Vukelich, eds. 1997. *Teaching language and literacy.* New York: Longman.

Echevarria, J., M. Vogt, and D. Short. 2004. *Making content comprehensible for English learners: The SIOP Model,* 2nd ed. Boston, Pearson.

Fathman, A., and M. Quinn. 1989. *Science for language learners.* Englewood Cliffs, NJ: Prentice Hall Regents.

Florida Department of Education. 1996a. *Florida curriculum framework: Language arts—preK–12 Sunshine State standards and instructional practices.* Tallahassee, FL: Author. Available at http://sunshinestatestandards.net.

Florida Department of Education. 1996b. *Florida curriculum framework: Science—preK–12 Sunshine State standards and instructional practices.* Tallahassee, FL: Author. Available at *http://sunshinestatestandards.net.*

Freeman, D., I. Freeman, M. McCloskey, L. Stack, C. Silva, A. Garcia, and M. Gottlieb. 2003. *On our way to English.* Barrington, IL: Rigby.

Gibbons, P. 2002. *Scaffolding language, scaffolding learning: Teaching second language learners in the mainstream classroom.* Portsmouth, NH: Heinnemann.

Haley, M., and T. Austin. 2004. *Content-based second language learning: An interactive approach.* Boston: Pearson.

Hill, J., C. Little, and J. Sims. 2004. *Integrating English language learners in the science classroom.* Allston, MA: Trifolium Books.

Kosek, J. 1999. *What's inside the Sun?* New York: Rosen Publishing Group.

National Research Council. 1996. *National Science Education Standards.* Washington, DC: National Academy Press. Online version at *www.nap.edu/books/0309053269/html/index.html.*

Schifine, A., D. Short, J. Villamil Tinajero, E. Garcia, E. Hamayan, and L. Kratky. 2004. *Avenues.* Carmel, CA: Hampton Brown.

Shin, F. 2005. *ELD in the content area: Science.* New York: Rosen.

Silverstein, S. 1964. *The giving tree.* New York: HarperCollins.

Texas Education Agency. 1998. *Texas state science standards.* Austin, TX: Author. Available at *http://www.tea.state.tx.us/teks.*

Further Reading

Chamot, A., and J. O'Malley. 1994. *The CALLA Handbook: Implementing the cognitive academic language learning approach.* Reading, MA: Longman. The authors present various aspects of lesson development that are considered appropriate for English learners. These are presented separately for skills that are needed for science learning as well as language skills specifically required by science for greater comprehension.

Curtain, H., and C. Dahlberg. 2004. *Languages and children—Making the match: New languages for young learners.* Boston, MA: Pearson. The authors make a case for using thematic instruction for curricular, unit, and lesson design. There is a step-by-step section about how to construct a lesson for daily implementation along with many considerations that teachers must entertain as they think about their students and what will benefit them most in any given lesson or topic.

Hill, J., C. Little, and J. Sims. 2004. *Integrating English language learners in the science classroom.* Allston, MA: Trifolium Books. This resource is a wonderful compendium of hands-on science lessons that integrate English language skills. Organized for easy reference in broad applications for lessons both in and outside the classroom, this volume is exceptionally teacher friendly and useful.

Marzano, R., D. Pickering, and J. Pollock. 2001. *Classroom instruction that works: Research-based strategies for increasing student achievement.* Alexandria, VA: Association for Supervision and Curriculum Development. The authors address classroom practice from various perspectives including placing it in the context of identifying similarities and differences. They also incorporate suggestions for making classroom practice effective by providing examples, of graphic organizers, note-taking skills and cooperative learning.

Section IV

Contexts for Classroom Implementation

Standards for Science and English Language Proficiency

Margo Gottlieb and Norman G. Lederman

At an urban middle school, an academic core team of teachers has just finished teaching the life science block to the seventh-graders and now is beginning preparations for the physical science block. The science teachers have taught this individually many times, but are looking a bit more carefully at the standards, not only for the eighth-grade performance test but also to prepare their students for high school courses.

This school had gone to block scheduling several years previously, and these teachers are part of an academic core team. They plan thematic instruction that is shared by the teachers of English, social studies, science, mathematics, and English as a second language (ESL) when possible. As the team meets to plan, they all have thick copies of their state academic content and English language proficiency standards. They discuss a possible theme of energy, specifically force and motion or potential and kinetic energy.

The social studies teacher says she can do a historical section on the use of energy throughout the Industrial Era in the United States. The English teacher says he has specific standards to cover, all of which deal with writing. He suggests that he could help students in writing a research report. The English as a second language or bilingual teacher then explains how many of the language proficiency standards could be covered while doing all of these different activities. Because the ESL/bilingual teacher has students at different English proficiency levels, she can help them using various instructional strategies. All the teachers have their own content standards to cover, but, with team planning, they determine where standards overlap and can be integrated into thematic instruction, making teaching easier and learning more cohesive for the students.

Window *Into the* Classroom

Standards in science from the National Research Council (NRC) advocated by the National Science Teachers Association (NSTA) and most other science education organizations and standards for English language learners (ELLs) advocated by Teachers of English to Speakers of Other Languages (TESOL) have until recently been developed independently. As stipulated under the No Child Left Behind Act, however, language standards for ELLs should be anchored in the academic content standards of science—in addition to language arts and mathematics.

Work is being done at national, state, and local levels to integrate these standards. As we are all well aware, local standards must take precedence over those developed at the national level. Nevertheless, the national standards have provided significant guidance for those developed at the state and local levels. Consequently, teachers and administrators can benefit significantly from a careful reading of the national standards.

At the national level, English language proficiency standards are being developed that include the language of scientific inquiry, creating a continuum that delineates the progression of acquiring the language of science along with the skills and knowledge of science. Although content areas still have their own standards, the new language proficiency standards for ELL students will blend science and language in an integrated approach similar to that described in the academic core team planning of the "Window Into the Classroom."

This chapter discusses aspects of sci-ence and English standards for language learners, in particular, their historical backdrop, organizing scheme, and integrated use. It will first cover the development and content of the National Science Education Standards, then the development and content of the national TESOL standards, and finally an initiative aimed at combining these two sets of standards in the development of new integrated English language proficiency standards.

Science Education Standards

In 1985, the American Association for the Advancement of Science (AAAS) quietly began a long-term initiative to reform K–12 education in natural and social science, mathematics, and technology. The initial product of this effort was *Science for All Americans* (Rutherford and Ahlgren 1989), a vision statement of a reform effort more popularly known as Project 2061. One year later, the National Council of Teachers of Mathematics (NCTM) began its effort to delineate national standards in mathematics curricula, teaching, and assessment for K–12 levels. The culmination of this effort was the publication of NCTM's *Curriculum and Evaluation Standards* (NCTM 1989). Although the work on Project 2061 preceded that of the mathematics community, the NCTM standards are often credited with beginning the flurry of standards development projects currently seen at the national, state, and local levels.

The reforms have much in common, but they are significantly different in some areas. The following discussion of the national reform documents in science education is presented in chronological sequence and does not represent an ordering by impact or preference. At the same time that these reform efforts were taking place in science, similar efforts were taking place in language arts and English as a second language.

Science Curriculum Reforms: Project 2061 and SS&C

After three years of planning, Project 2061 was launched in 1985 under the direction of F. James Rutherford. Project 2061 is based on the belief that the K–12 education system should be reformed so that all American high school graduates are science literate (i.e., students will possess the knowledge and skills to lead interesting, responsible, and productive lives in a culture and society increasingly influenced by science and technology).

Six assertions guide Project 2061:

1. Reform must be comprehensive and center on all children, grades, and subjects and must represent a long-term commitment.

2. Curriculum reform should be dictated by our collective vision of the lasting knowledge and skills needed by our students and future citizens.

3. The common core of learning in science, mathematics, and technology should focus on science literacy as opposed to preparation of students for careers in science. The core curriculum should emphasize connections among the natural and social sciences, mathematics, and technology. The connections with these areas and the arts, humanities, and vocational subjects should be clear as well.

4. Schools should teach less and teach it better. Superficial coverage of specialized terms and algorithms should be eliminated, an example of the often-quoted phrase, "less is more."

5. Reform should promote equity in science, mathematics, and technology education, serving all students equally well.

6. Reform should allow more flexibility for organizing instruction than is currently common.

The initial, and most influential documents produced by Project 2061 were *Science for All Americans*, which presents a vision and goals for science literacy, and *Benchmarks for Science Literacy*, which translates the vision of *Science for All Americans* into expectations, i.e., benchmarks, for core content by the end of grades 2, 5, 8, and 12. Since the publication of these books, *Designs for Science Literacy*, *Curriculum Materials Resource*, *Resources for Science Literacy: Professional Development*, *Blueprints for Reform*, and the *Atlas of Science Literacy* have been published. They are designed to help teacher educators prepare teachers in a manner consistent with the goals of Project 2061 and to assist practicing teachers in the implementation of project goals in their classrooms.

The Scope, Sequence, and Coordination of Secondary School Science (SS&C) (1989) reform initiative advocated coherent, carefully sequenced science curriculum that is thematic and interdisciplinary. SS&C had four publications, with the most significant being *Scope, Sequence, and Coordination of Secondary School Science: Relevant Research* (NSTA 1992). The publication of *A High School Framework for National Science Education Standards* (Pearsal 1993), another SS&C publication, was the first in a series that led to NSTA's shift toward the adoption of the National Science Education

Standards. Full support for the NSES is now published in both the NSTA position statement (1998; online at *www.nsta.org/position-statementandpsid=24*) and the series of Pathways to the Science Education Standards (Texley and Wild 2004).

Many state science teachers association have taken similar actions—Oregon, Illinois, and New York, to name just a few. Overall, SS&C advocates science programs that help students answer questions such as, "How do we know?," "Why do we believe?," and "What does it mean?"

National Science Education Standards

In January 1996 the National Research Council of the National Academy of Sciences (NRC) published the *National Science Education Standards* (NSES).

The project is a more comprehensive endeavor than either Project 2061 or SS&C, as it not only provides a vision for scientific literacy but also a framework for how the vision should be realized. It is important to note that the primary overlap of purpose between *NSES* and *Benchmarks for Science Literacy* lies in the subject matter standards and benchmarks. The NSES advocate that all citizens be able to develop and conduct a scientific investigation while Project 2061 simply advocates that our citizens be able to understand the various aspects of a scientific investigation. Although *NSES* can be viewed as the more ambitious document, one can also argue that, relative to the goal of scientific literacy, an understanding about scientific inquiry is far more valuable and realistic than the ability to conceive, design, and carry out a scientific investigation. Nevertheless, most educators would agree that the best way to teach about inquiry is to have students do inquiry and then reflect on what they have done.

The goals for school science specified in *NSES* are to educate students who are able to

1. use scientific principles and processes appropriately in making personal decisions,
2. experience the richness and excitement of knowing about and understanding the natural world,
3. increase their economic productivity, and
4. engage intelligently in public discourse and debate about matters of scientific and technological concern.

NSES is based on four basic principles, (Table 1), decreased from the seven in the 1994 draft:

The six sets of standards delineated in

Table 1: Four principles of the National Science Education Standards

1. Science is for all students.

2. Learning science is an active process.

3. School science reflects the intellectual and cultural traditions that characterize the practice of contemporary science.

4. Improving science education is part of systemic education reform.

NSES comprise the majority of the published document:

- Science Teaching Standards,
- Standards for Professional Development,
- Assessment in Science Education,
- Science Content Standards,
- Science Education Program Standards, and
- Science Education System Standards

School systems across the nation, as well as professional organizations, have recently focused on the NSES as a reference point, but that focus is somewhat diffuse as it relates to individual classrooms and practice. Consequently, the subsequent discussion on the implications of reform for teaching and learning will use the NSES as a reference point.

The Standards for the Teaching of Science clearly outline how instruction should be revised so that it models scientific inquiry. In particular, these Standards elaborate on how a teacher should design lessons so that students are placed in situations that require them to collect or analyze data and arrive at inferences or conclusions concerning the meaning of the data. The teaching Standards emphasize student-centered instruction, with the ideal being situations in which students develop questions of interest and ways to answer the questions and then carry out the investigation.

Interestingly, the Standards for Professional Development parallel the Standards for teaching in that the teacher is also viewed as a learner who enters every learning situation (in this case professional development) with background knowledge and predispositions. As a consequence, the professional development Standards move away from a "one-size-fits-all" mentality and from the artificial distinction between preservice and inservice teachers.

The Standards for Sssessment clearly recognize that students will represent what they have learned in ways as diverse as the ways in which they are known to have learned. The Standards also recognize that the types of in-depth knowledge stressed in the content Standards are not easily assessed, in general, by traditional paper-and-pencil tests. Consequently, these Standards stress a variety of assessment techniques such as performance tasks, portfolios, and samples of thought in addition to traditional assessment.

The content Standards do not differ much from other lists of traditional content, except that they evidence an effort to focus more on overarching themes. Perhaps the most noticeable feature of the content Standards is their emphasis on the nature of science and inquiry (doing as well as knowing). The doing of inquiry is not new to science teachers. We have had students collect data, make measurements, and draw inferences for many years. The Standards, however, also stress that students should know about inquiry. This involves learning concepts such as there is no single scientific method, all science does not necessarily involve an experiment, and different scientists may come to different conclusions even though they completed the same procedures.

Although all the NSES are used in various aspects of science education and teaching, classroom teachers rely on the Science Content Standards section on a daily basis for instruction. The content Standards consist of seven topical areas organized into K–4, 5–8, and 9–12 grade levels. These

Table 2: NSES Science Content strands

- Science as inquiry
- Physical science
- Life science
- Earth and space science
- Science and technology
- Science in personal and social perspectives
- History and nature of science

seven areas—which are not listed in any order of importance—are seen in Table 2.

The specific outcomes for each topical area are organized into a logical and developmentally appropriate sequence with increasing complexity as one moves to higher grade levels. For example, within life science, the scope and sequence of topics appears as follows:

K–4

Characteristics of organisms
Life cycles of organisms
Organisms and environment

5–8

Structure and function in living systems
Reproduction and heredity
Regulation and behavior
Populations and ecosystems
Diversity and adaptation of organisms

9–12

The cell
Molecular basis of heredity
Biological evolution

Interdependence of organisms
Matter, energy, and organization in living systems
Behavior of organisms

Although most individuals focus on subject matter when considering the NSES, subject matter outcomes represent only one portion of the NSES. More detailed information is available at the National Research Council (NRC) website for the National Science Education Standards at *www.nap.edu/ readingroom/books/nses/html*.

NSES is very clear in stating that the Standards, especially the Content Standards, are not meant to be a curriculum. Different states and school systems define their own curricula so that school offerings can be customized to local needs and interests. For example, inland states like Illinois have chosen to place less emphasis on marine environments than does Oregon. Illinois also places less emphasis on volcanoes than does Oregon. The Nevada state standards place a stress on mining that is not evident in the standards of Illinois or Oregon. The message is clear: the NSES Content Standards are a framework from which to design curriculum, but they also allow flexibility in the specific content emphasized from one state to another.

The Standards on science programs and systems are designed to equip teachers and administrators with a rationale for reform and the system and program needs to accomplish the visions of the NSES. The overall message is that science education reform is not the responsibility of one group of individuals but rather is the result of a joint effort by numerous constituencies in the school and community. In order to accomplish reform, teachers, administrators, and community need a common vision, and the necessary resources must be provided. If we want teaching and learning to be inquiry oriented, *NSES* states, then teachers have to value inquiry, appropriate professional development must be provided, assessment must be reorganized, and administrators must value inquiry and help provide the support teachers need to make the shift to inquiry instruction.

Implications for Science Teaching

Clearly much can be said about reform in science education and what needs to be done to achieve the visions of the reforms. The rhetoric can be overwhelming for many teachers, and it can create a less-than-healthy skepticism in others. In the end, the message that should resonate with most teachers is "science as inquiry." This phrase encompasses the idea that science should be taught using an inquiry-oriented approach, that students should be expected to learn how to do science as well as learn about inquiry.

On the teaching side of inquiry, the assumption is that students will more likely learn science subject matter if they learn it in the same way that scientists learn subject matter.

On the learning side, we want our students to learn how to do inquiry and to learn about inquiry, as well as learn the traditional subject matter. The rationale for this expansion of learning outcomes is related to the current stress on scientific literacy. If students and the general public are to make informed decisions about scientific claims and scientifically based societal and personal issues, they need to know how scientific knowledge is devel-

oped and they need to know about the limitations of the knowledge. An understanding of how scientific knowledge is developed readily reveals why scientific knowledge is never absolute and how it is possible for two scientists to disagree. The scientists do not disagree because one is wrong and one is right. They disagree because they are reviewing and interpreting the existing data from differing perspectives.

Many of the authors in this book have discussed support for inquiry-based instruction: in Chapter 3 for planning lessons, in Chapter 4 for classroom activities, and in Chapter 7 for designing lessons. And, there is more than emerging data to support that using inquiry-oriented teaching approaches can lead to students' understanding of both inquiry and traditional subject matter. In addition, other potential practical outcomes are associated with an inquiry-oriented teaching approach. To name just two, an inquiry-oriented approach can be used to facilitate integration of subject matter and it can be used to more reasonably meet the needs of all students, including language learners, as described in the "Window Into the Classroom" at the beginning of this chapter.

During the past decade, the arbitrary distinction schools make among various disciplines has been much discussed. Science teachers have recognized that scientifically based problems are not the purview of a particular area of science or science in general. For example, is the problem of increasing pollution in Lake Michigan simply a scientific problem to be solved by scientists? Clearly not. The problem is a problem for the Chicago community and all communities that border on Lake Michigan. The solution to the problem involves areas of science, political science, sociology, mathematics, and engineering. In short, if curriculum and instruction are organized around authentic inquiry, students will be attempting to solve problems that involve several disciplines. We can think of no better way than inquiry-based teaching to integrate the school curriculum so that what students experience is relevant to their everyday lives.

Much of the research on group learning and cooperative learning, as already noted in Chapter 2 and Chapter 5, has shown that students with diverse abilities, backgrounds, and cultures increase their learning within such an instructional goal structure or organization.

In an inquiry-oriented classroom, students with diverse backgrounds—including speakers of other languages—must work together toward the solution to scientific problems or toward a more descriptive investigation. No matter the topic, having students work together as they do during scientific inquiry has the potential to not only improve achievement but also improve student affect, language skills, and interpersonal relationships. The diversity of students in a typical classroom is continually increasing, and the challenge this poses for the classroom teacher is clear.

Using an inquiry-oriented teaching approach appears to be a promising response to the challenge. "Promising" here means to communicate a solution that enhances the cognitive, social, and emotional development of all students. "Science for all" has been a mantra of science education reform efforts. "Science as inquiry" includes the notion that all students can do, learn, and appreciate science.

English Language Proficiency Standards

In the early 1990s, national standards were being developed in content areas such as science, history, English language arts, and geography, but, at that time, most content-area standards did not take into account the needs of English language learners. In response to the unique characteristics of linguistically and culturally diverse students, Teachers of English to Speakers of Other Languages (TESOL), the international organization of English language educators, began formulating English as a second language (ESL) standards. The *Access Brochure* (TESOL 1993) was developed by a task force of ESL experts that encouraged the inclusion of English as a second language in the ongoing educational reform movement. In 1997, TESOL published *ESL Standards for Pre-K–12 Students*. These standards described what constitutes effective education for students acquiring English. They were not intended to replace standards in other content areas, but to supplement them (TESOL 1997).

The ESL standards included three goals with nine standards that described what students should know and be able to do using English as a result of instruction, as shown in Table 3.

The standards were organized into grade-level clusters (Pre-K–3, 4–8, and 9–12). Each cluster addressed all goals and standards, followed by descriptors (behaviors students demonstrate when they meet a standard), progress indicators (observable activities that students may perform to show progress toward meeting the standard), and vignettes (instructional sequences that show the standard in action). A description of these elements is shown in Table 4.

Table 3: English as a Second Language Standards for Pre-K–12 Students (TESOL 1997)

ESL STANDARDS FOR PRE-K–12 STUDENTS

Goal 1: To use English to communicate in social settings
Standard 1: Students will use English to participate in social interaction.
Standard 2: Students will interact in, through, and with spoken and written English for personal expression and enjoyment.
Standard 3: Students will use learning strategies to extend their communicative competence.

Goal 2: To use English to achieve academically in all content areas
Standard 1: Students will use English to interact in the classroom.
Standard 2: Students will use English to obtain, process, construct, and provide subject matter information in spoken and written form.
Standard 3: Students will use appropriate learning strategies to construct and apply academic knowledge.

Goal 3: To use English in socially and culturally appropriate ways
Standard 1: Students will use the appropriate language variety, register, and genre according to audience, purpose, and setting.
Standard 2: Students will use nonverbal communication appropriate to audience, purpose, and setting.
Standard 3: Students will use appropriate learning strategies to extend their sociolinguistic and sociocultural competence.

Table 4: Sample goal and standard from ESL standards (TESOL 1997)

Goal 2, Standard 1 in Grades 4–8: To use English to achieve academically in all content areas: students will use English to interact in the classroom.

Sample descriptor: Students can request information and assistance.

Sample progress indicator: Students request supplies to complete an assignment.

Sample vignette: A description of a sheltered science class where students examine containers of various shapes and hypothesize which containers have more or less liquid and then evaluate predictions by measuring the capacity of each container.

The national English as a Second Language Standards have been used in the development of state English as a second language (ESL) or English language development (ELD) standards. ESL standards put the educational spotlight on students' acquiring ESL and explained how educators can help these students move successfully from ESL or bilingual classrooms to mainstream classrooms (Gómez 2000). It also has provided the backdrop for recent standards revisions that focus on developing language standards with integrated content.

English Language Standards Integrating Language and Content

For the education community serving English language learners, the notion of developing and implementing English language proficiency standards that connect with academic content standards is a recent phenomenon. This effort has been spurred by three converging situations:

* a shift in pedagogical practice,
* a call for a new generation of language proficiency assessments, and
* educational policies arising from the No Child Left Behind Act of 2001 (Gottlieb 2003).

As a result, individual or consortia of states are outlining, through English language proficiency standards, what constitutes the language necessary for social and academic success in school.

This vision is synchronized with theory and content-based instructional techniques for English language learners that have been prevalent for the last couple of decades (from Mohan 1986 to, most recently, Echevarria, Vogt, and Short 2004). It is also

in concert with a continuous plea, over the same 20-year span (from Cummins 1981 to Bailey and Butler 2002), for academic language proficiency assessment that captures the language necessary for mainstream classroom participation.

The development of ESL or ELD standards prior to 2001 by TESOL and some states is important. After the No Child Left Behind Act of 2001, however, language proficiency standards for English language learners should move beyond language arts and include the academic language of mathematics and science, as demonstrated in Chapter 3, which discusses the integration of standards in planning instruction. Parallel sets of standards for English language proficiency and academic content give English language learners enhanced opportunities for academic success. The ultimate goal is to have English language learners attain proficiency on both sets of standards as represented in Figure 1 (TESOL 2005).

In the past, teachers working exclusively with English language learners have been isolated, both physically and psychologically. ESL and bilingual teachers often have felt that the sole responsibility for the education of these students rests on their shoulders. With the introduction of English language proficiency standards that clearly mark the pathways to acquisition of English through science (as well as the other core content areas), teaching becomes a more collaborative, unified endeavor. As science topics are embedded within English language proficiency standards, ESL/bilingual and general education teachers can more readily coordinate delivery of a common science curriculum by jointly planning science lessons, experi-

Figure I: TESOL English Language Proficiency Standards in Core Content Areas

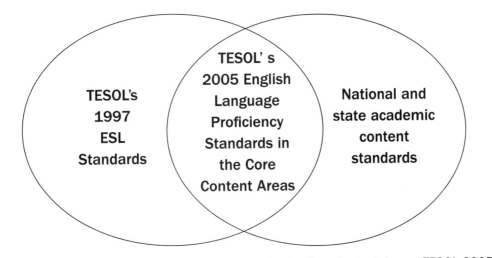

Pre- K–12 English Language Proficiency Standards in the Core Content Areas (TESOL 2005) Draft copy.

ments, and field experiences.

The Development and Organization of English Language Proficiency Standards

Crafting English language proficiency standards, which integrate content and language standards, is arduous, labor-intensive work that involves input from and the consensus of multiple stakeholders. Each state or consortium of states has approached this task along different lines. The format for the standards outlined in this section is the product of the World-Class Instructional Design and Assessments (WIDA) Consortium of 10 states that, in turn, has become the prototype for TESOL's new pre-K–12 English Language Proficiency Standards, scheduled for publication in 2006 (Gottlieb 2004).

The content areas of language arts, science, mathematics, and social studies are incorporated into the English Language Profi-

ciency Standards in the Core Content Areas. Teachers and administrators involved in conceptualizing these standards want to highlight the importance of sustained content-driven instruction as an entrée to, and preparation for, rigorous state academic standards. Table 5 lists English language proficiency standards that are being adopted by TESOL (State of Wisconsin 2004).

TESOL, in its preview document for its pre-K–12 English Language Proficiency Standards (2005), has added the intercultural dimension of language learning. One application of sociocultural competence in a science classroom is the student's appropriate use of formality in writing a lab report versus the informality of communicating with peers when conducting scientific investigations

Under Title III of the No Child Left Behind Act, each state has latitude in organizing its English language proficiency standards,

Table 5: English language proficiency standards

PreK–12 English Language Proficiency Standards

1. English language learners communicate in English for social and instructional purposes within the school setting.

2. English language learners communicate information, ideas, and concepts necessary for academic success in the content area of language arts.

3. English language learners communicate information, ideas, and concepts necessary for academic success in the content area of mathematics.

4. English language learners communicate information, ideas, and concepts necessary for academic success in the content area of science.

5. English language learners communicate information, ideas, and concepts necessary for academic success in the content area of social studies.

---

as long as it complies with the basic requirements stipulated under the law. The WIDA Consortium has chosen to arrange its five standards in the following configuration.

The "Classroom Framework" describes how teachers can teach and assess English language learners' use of the language of science on an ongoing, and perhaps extended, basis by building models, engaging in experimentation, and using equipment, instruments, and technology as part of inquiry.

The "Large-Scale Assessment Framework" measures those aspects of science that are amenable to short-term teaching and testing, such as completing tables, charts, and graphic organizers, and classifying or categorizing scientific information.

Four language domains define how English language learners process, understand, interpret, and evaluate receptive language (through listening and reading) as well as en-

gaging in productive communication in a variety of situations for different audiences and purposes (through speaking and writing).

Four grade level clusters reflect the developmental progression of English language learners as they move through school: K–2, 3–5, 6–8, and 9–12.

Five levels of English language proficiency identify the sequential stages of second language acquisition: Entering, Beginning, Developing, Expanding, and Bridging.

Model performance indicators, which are formed by the intersection of grade level clusters with levels of English language proficiency, specify how English language learners use language associated with topics named within the content areas. Common science topics have been gleaned from a content analysis of science content standards of the affiliated states that, in most instances, are derivations of the National Science Education Standards.

In addition to these basic elements or features, there are performance definitions that overarch the standards. These broad statements describe the expectations of English language learners at each level of language proficiency, and they provide additional information to assist classroom teachers in differentiation of instruction and assessment.

Integration of Language and Content Using Examples From Science

As a federal requirement since 2003, state English language proficiency standards delineate the relationship between acquisition of language and content for English language learners. The integration of language and content, in this case in the area of science, occurs within and across standards. First, within the English language proficien-

cy standards, the juxtaposition of language and content is visible within the model performance indicators themselves. The series or strands of model performance indicators also form a developmental continuum of progressive English language proficiency levels. Table 6 illustrates how the English language proficiency standard for science represents the interaction between language and content within its model performance indicators (State of Wisconsin 2004).

Each model performance indicator for the science standard consists of a language function, a science topic, and the type of support necessary for English language learners at the designated language proficiency level. In this example, the science topic, natural disasters, is representative of the NSES Content Standard F, "as a result of activities in grades 5–8, all students should develop understanding of natural hazards." It provides the content thread

Table 6: An example of the interaction between language and content for the five levels of English language proficiency for Standard 4

Grade Level Cluster: 6–8	Language Domain: Reading
Classroom Framework	

Language Proficiency Level	Model Performance Indicator
5—Bridging	Interpret impact of natural disasters on people and places from grade level text.
4—Expanding	Compare natural disasters using multiple written sources, including the internet and graphic organizers.
3—Developing	Identify characteristics and conditions related to natural disasters based on text and pictures.
2—Beginning	Respond to questions regarding natural disasters based on graphic organizers and pictures.
1—Entering	Chart time and places of natural disasters based on headlines and pictures.

throughout the strand of the five English language proficiency levels.

Although the topic remains constant throughout the strand as can be seen in Table 6, the language functions, or how English language learners use language, changes as their English language proficiency becomes more fully developed. There is a naturally occurring blend of language and content within the model performance indicators.

English language proficiency standards are to be anchored in state academic content standards. This alignment creates a seamless pathway for English language learners to cross from language proficiency to academic achievement. The matching of English language proficiency with academic content exemplifies how language and content can be integrated across standards. Table 7 shows how a specific science concept (extracted from the National Science Education Standards) links English language proficiency with academic content standards.

Implications for Teaching

Every discipline has a language unto itself, and this principle applies to science as well. For English language learners to thrive in general education classes—the goal of most support services for these students—they need to be exposed to content-based language in natural settings. As noted previously in this chapter, an inquiry approach to science, in which all students have opportunities to investigate firsthand, facilitates the simultaneous development of science and concepts (Kessler and Quinn 1987; Mohan 1986).

English language proficiency standards, and their accompanying performance or progress indicators, have to be broad enough to honor the linguistic and cultural diversity of English language learners. Unambiguous

Table 7: Integrating English language proficiency and academic content standards through concepts

English Language Proficiency Standard	Science Concept	Academic Content Standard
English language learners communicate information, ideas, and concepts necessary for academic success in the content area of science	Internal and external processes of the Earth system cause natural hazards, events that change or destroy human and wildlife habitats, damage property, and harm or kill humans.	Science in Personal and Social Perspectives
Language Level 5: Interpret the impact of natural disasters on people and places		All students should develop understanding of natural hazards

statements saying which linguistic patterns and technical language associated with science concepts are appropriate at each stage of language development provide teachers with a map for instruction and assessment. In turn, teachers must take their students' levels of English language proficiency, as well as the students' academic achievement and previous educational experience, into account in their lesson design and delivery.

Another consideration in teaching and assessing the language of science that should be implicit in language proficiency standards is the cultural backgrounds of English language learners. These students may not be accustomed, for example, to the climatic conditions within the United States, or may have different types and sources of food. In which food group does the chirimoya, indigenous to Chile, belong? In addition, students may have varying perspectives of how science is organized; for example, Native Americans may categorize plants and animals according to their function rather then structure (Smith 1986).

Because inquiry is central to the teaching of science, teachers of English language learners must be sensitive to expected student performance at varying English language proficiency levels. Think about how students develop questions, for example, the starting point for all scientific investigations. Contingent on their English language proficiency, some students may be asked to produce simple phrases or statements that are visually supported; others, to restate or summarize a prediction in their own words; and still others, to generate original questions. Older students may be asked to produce scientific language to form a hypothesis and, later, to test the hypothesis. These language patterns need to be introduced and reinforced for English language learners before moving to the next step in the sequence.

An Integrated Approach to Standards-Based Instruction and Assessment

English language proficiency standards help shape curriculum, instruction, and assessment for English language learners. Those that address the language of science guide content-based instruction in science where, as we have seen, concepts are introduced with visual or graphic support while linguistic complexity is reduced. However, acquiring English language proficiency is only half of the equation for these students.

At the same time they learn English, English language learners must be gaining the knowledge and skills of science, either in English or their native language, as part of their academic achievement. In classrooms where English is the medium of instruction and teachers have been trained in ESL methodologies, or in dual-language classrooms in which there is native language support, science concepts form a bridge from English language proficiency to academic content standards so that language and content become integrated.

Collaborating among teachers and coordinating support services for English language learners are key to ensuring an integrated approach to standards-based instruction. Teachers, irrespective of the language of instruction, should have a shared vision of the expectations for English language learners and how the students can successfully access the curriculum. Standards provide the platform for accomplishing this goal.

In general, ESL and bilingual teachers preview science concepts, language patterns, and vocabulary specified in language proficiency standards while general education teachers extend language proficiency into academic achievement by overlaying the knowledge and skills of science from academic content standards. Easy to say, but how can we meet this challenge? Using the example of scientific inquiry, during an ESL lesson, teachers would concentrate on students' formulation of statements and questions using science vocabulary or the transition words that mark a sequence (first, next, finally). During a science lesson, teachers would center on whether English language learners could actually conduct an investigation or experiment to answer questions of interest or to collect data relative to the testing of a hypothesis.

Ongoing professional development and joint time for planning are essential for teachers to develop rich and deep standards-based instruction and assessment activities, tasks, and projects. Opportunities to combine multiple standards, and their performance indicators, to create thematic units of study are advantageous for English language learners (Freeman and Freeman 2002; Enright and McCloskey 1988). By capitalizing on the respective expertise of ESL/bilingual and general education teachers, all students can benefit from effective science instruction.

Applications for the Classroom

Standards are a resource and guide for administrators and teachers. For educators working with English language learners, standards provide a common yardstick that systematically defines language development as related to science, science process skills, and content knowledge.

In preparing and implementing standards-based instruction and assessment, teachers need a sense of the functions, or how students use language to communicate scientific ideas, and the contexts, or science topics, in which functions operate. Both of these elements are derived directly from English language proficiency standards. English language proficiency standards align naturally with science standards as a guide for assessment and instruction; this continuity is essential for developing a strong, unified curriculum for English language learners.

Additionally, teachers of English language learners must be cognizant of the interplay among the four language domains (listening, speaking, reading and writing) specified in English language proficiency standards and those domains' roles in language development. Table 8 is a suggested list of typical language functions and major science topics; it represents the findings from an analysis of English language proficiency and science content standards from 10 states. Teachers are welcome to interchange functions across the language domains and combine them with grade-specific topics as a starting point for developing science curriculum for their English language learners.

Some of the prevalent topics selected for the English language proficiency standard for science correspond to unifying concepts and processes in science. These include the crosscutting themes of systems and change that have applicability across scientific disciplines. As there is only one English language proficiency standard for science

Table 8: A sampling of language functions by language domain and science topics from English language proficiency standards

Language Functions by Language Domains			
Listening	**Speaking**	**Reading**	**Writing**
• Identify • Classify/sort • Categorize • Sequence • Role play or simulate	• Re/state • Ask and answer questions • Describe • Discuss • Debate	• Match • Interpret • Make inferences • Draw conclusions • Apply information	• Label/ Illustrate • Hypothesize • Compare/Contrast • Chart information or data • Summarize processes or findings
Prevalent Science Topics Across Language Domains			
¤ Change ¤ Cycles ¤ Systems	¤ Processes ¤ Light, sound, and heat ¤ States of matter	¤ Organisms ¤ Discovery and Invention ¤ Energy	¤ Natural resources ¤ Plants and animals ¤ Weather

and eight major science themes, the same treatment or representation cannot be expected to be present in each. For that reason, *model* is used with *performance indicators* for English language proficiency. Teachers and administrators working with English language learners can use these curricular kernels as a starting point for developing additional language proficiency strands to meet their local needs.

Summary

This chapter describes how standards for science and English language proficiency can be blended with science standards to create a continuum of language development and academic achievement for English language learners. The pair of standards works in tandem to create a unified, transparent crosswalk for ELLs as they acquire English. Equally important, the standards

are integrated so that teachers can collaborate to offer English language learners a cohesive curriculum, sound assessment, and effective instruction in science as well as other content areas.

References

American Association for the Advancement of Science. 1993. *Benchmarks for science literacy.* New York: Oxford University Press.

Bailey, A., and F. Butler. 2002. *An evidentiary framework for operationalizing academic language for broad application to K–12 education: A design document.* Los Angeles: University of California, Los Angeles.

Cummins, J. 1981. The role of primary language development in promoting educational success for language minority students. In *Schooling and language minority students: A theoretical framework,* ed. California State Department of Education, 3–49. Los Angeles: California State University.

Echevarria, J., D. Short, and M. Vogt. 2004. *Making*

content comprehensible for English language learners: The SIOP Model, 2nd ed. Boston: Allyn and Bacon.

Enright, D., and M. McCloskey. 1988. *Integrating English: Developing English language and literacy in the multilingual classroom.* Reading, MA: Addison-Wesley.

Freeman, Y., and D. Freeman. 2002. *Closing the achievement gap: How to reach limited-formal-schooling and long-term English learners.* Portsmouth, NH: Heinemann.

Gómez, E. L. 2000. A history of the ESL standards for pre-K–12 students. In *Implementing the ESL standards for pre-k–12 students through teacher education,* ed. M. A. Snow, 49–74. Alexandria, VA: Teachers of English to Speakers of Other Languages.

Gottlieb, M. 2003. Large-scale assessment of English language learners: Addressing accountability in K–12 settings. *TESOL Professional Papers #6.* Alexandria, VA: Teachers of English to Speakers of Other Languages.

Gottlieb, M. 2004. *WIDA consortium English language proficiency standards for English language learners in kindergarten through grade 12: Overview document.* Madison, WI: State of Wisconsin.

Kessler, C., and M. Quinn. 1987. ESL and science learning. In *ESL through content-area instruction,* ed. J. Crandall. Englewood Cliffs, NJ: Prentice Hall Regents.

Mohan, B. 1986. *Language and content.* Reading, MA: Addison-Wesley.

National Council of Teachers of Mathematics. 1989. *Curriculum and evaluation standards for school mathematics.* Reston, VA: Author

National Research Council. 1996. *National Science Education Standards.* Washington, DC: National Academy Press.

National Science Teachers Association. 1992. *Scope, sequence, and coordination of secondary school science: Relevant research.* Washington, DC: Author.

Pearsall, M., ed. 1993. *Scope, sequence, and coordination of secondary school science: The content core.* Washington, DC: National Science Teachers Association.

Rutherford, F. J., and A. Ahlgren. 1989, 1993. *Science for all Americans.* New York: Oxford University Press.

Smith, M. 1986. A model for teaching native oriented science. In *Science education and cultural environments in the Americas,* eds. J. J. Gallagher and G. Dawson. Washington, DC: National Science Teachers Association.

State of Wisconsin. 2004. *WIDA Consortium English language proficiency standards for English language learners in kindergarten through grade 12.* Madison, WI: Author.

Teachers of English to Speakers of Other Languages. 1993. *Access brochure.* Alexandria, VA: Author.

Teachers of English to Speakers of Other Languages. 1997. *ESL standards for pre-K–12 students.* Alexandria, VA: Author.

Teachers of English to Speakers of Other Languages. 2005. *PreK–12 English language proficiency standards in the core content areas preview document.* Alexandria, VA: Author.

Texley, J., and A. Wild. 2004. NSTA pathways to the *science standards,* 2nd ed. Arlington, VA: NSTA Press.

Further Reading

American Association for the Advancement of Science. 1993. *Benchmarks for science literacy.* New York: Oxford University Press. A comprehensive listing of what students should know and be able to do covering grades K–12. This document is derived from *Science for all Americans* and it presents outcomes in a concrete form that is easily understood by educators.

Freeman, Y., and D. Freeman. 2002. *Closing the achievement gap: How to reach limited-formal-schooling and long-term English learners.* Portsmouth, NH:

Heinemann. This book stresses the importance of thematic instruction for English language learners as a means of intertwining language and content in a systematic and sustained way. There are vivid examples of how to scaffold lessons into units of instruction where individual language and academic needs are addressed for diverse learners.

Gottlieb, M. 2003. Large-scale assessment of English language learners: Addressing accountability in K–12 settings. *TESOL Professional Papers #6.* Alexandria, VA: TESOL. This monograph responds to the pressing need to develop assessments for English language learners that reflect effective educational theory and practice while meeting the requirements of the No Child Left Behind Act. The integration of language and content provides a useful model for defining and measuring academic language proficiency.

Rutherford, F., and A. Ahlgren. 1989, 1993. *Science for all Americans.* New York: Oxford University Press. This text presents the original vision for reform in science education that led to the development of specifications for student learning outcomes. An excellent source for the underlying rationale of current learning standards.

Perspectives on Teaching and Integrating English as a Second Language and Science

Deborah J. Short and Marlene Thier

Mr. Hafiz teaches physical sciences for the eighth grade. His class has spent the past two days doing hands-on inquiry activities that introduced basic simple machines and demonstrated how the machines work when a force is applied on them. His class is composed of 27 students of Hispanic descent. More than half are English language learners, ranging in proficiency from beginning to advanced. Other students are former ELLs, and some are native speakers of English.

As a review of the inquiry labs done previously, Mr. Hafiz shows the class pictures of three types of simple machines they have already studied—inclined plane, wedge, and lever—and asks each student to tell a partner the names of the images and give an example of how each machine is used in their daily lives. Students can interact while reviewing the material, and the less proficient students can speak to other students with less anxiety.

After the students discuss the simple machines, Mr. Hafiz selects three volunteers to share ideas with the class. He has the class create a flip book using three sheets of paper, folded and nested, to make six flippable pages. The students write the names of the three machines on the front of the top three pages and when they flip them up, they draw pictures and write the definition of how a simple machine operates in terms of work and force along with an example on the back side.

Mr. Hafiz then introduces the students to three additional simple machines—pulley, wheel and axle, and screw—with real objects and oral explanations. He divides the class into six groups and assigns each one of the new machines to two groups. He encourages the groups to try out their assigned machines at three stations he has set up. Then he asks each group to read about their assigned machines in the *textbook and respond to the following questions on a concept map: "What is it? What is it used for? What force is applied thus resulting in easier work? What are other examples of the machine?"*

After the groups read the assigned material, they discuss their machines. The beginning learners of English can read the Spanish version of the textbook if they prefer. The group discussions are in English with Spanish clarification. After the discussions conclude, each student answers the concept map questions in English, although Mr. Hafiz allows the beginners to copy notes from colleagues.

When all groups have finished, Mr. Hafiz reconfigures the students into triads; each triad has one student representing a different machine. In arranging these new groups, Mr. Hafiz tries to place a student with strong proficiency in English in each one. The students teach one another about their machines, and they complete the flip books with the new information. Mr. Hafiz wraps up the lesson with an "Outcome-Sentence" review of what students have learned about simple machines. Tossing a small stuffed ball around, students take turns completing sentence starters like "I learned ..., I wonder..., I still have a question about ..., and I discovered ...."

Many English language learners (ELLs) are struggling academically, as has been noted many times in this book. State and district leaders are worried about persistent, significant achievement gaps between ELLs and native English speakers on state and national assessments and the higher dropout rates among ELLs (California Department of Education 2004; Latinos in Education 1999; Snow and Biancarosa 2004). Science educators need to address the dramatic, enduring divide in science achievement between Caucasian students and those from culturally and linguistically diverse groups (Siegel 2002).

One reason for the achievement gap may be lack of teacher preparation. Despite a consistent appeal in science education literature for approaches that teach science effectively to all students (NRC 1996), most teachers have not had specific training in making content comprehensible and developing academic science literacy among ELLs. Unlike Mr. Hafiz, many teachers do not know how to adjust instruction for these students' second language development needs. As the National Commission on Teaching and America's Future has reported (1996), there are significant shortages of teachers qualified to teach students with limited English proficiency and of bilingual teachers trained to teach core subjects in another language.

Although the No Child Left Behind Act of 2001 calls for highly qualified teachers in every core academic classroom by 2006, neither the legislation nor most states' implementation plans require that the elementary classroom and secondary content area teachers have an educational background or training in second language acquisition theory, ESL methodology, or cross-cultural communication, despite the fact that the number of English learners is rising significantly throughout U.S. schools. In the 1999–2000 Schools and Staffing Survey (National Center for Education Statistics 2002), 41.2% of the 2,984,781 public school teachers reported teaching limited-English-proficient students but only 12.5% had had eight or more hours of training in the past three years. Few educators believe that eight hours is even a minimum amount of the professional development time needed to teach ELLs well.

In science, ELLs must cover a great deal of cognitively demanding material; read textbooks and comprehend abstract concepts; and conduct, orally report on, and write about experiments and other classroom tasks using a language they are still learning. Students need extra time to learn English and science. However, schools are under pressure to emphasize reading and mathematics due to high-stakes assessments, so science education has had to struggle to maintain visibility in the instructional day (Saul 2004). Many schools, particularly at the elementary level, have cut the amount of time they offer science lessons to students (Lee and Fradd 1998; Thier and Daviss 2001) and this does a disservice to the ELLs. Although this situation may shift in the 2006–07 school year (when the No Child Left Behind Act mandates that testing of science knowledge begin), science educators still need to learn about and implement successful, research-based interventions for teaching science to ELLs. Only then will the achievement gap between ELLs and native-English-speaking students be reduced.

In this chapter, we set the context for change. After briefly reviewing the history of ESL and science education, we discuss current promising practices that integrate ESL, literacy, and science education. We conclude with highlights of some programs being tested now in U.S. schools that may offer additional interventions for improving the science achievement of English language learners.

Changes in English as a Second Language Instruction

During the past century, the way we teach second languages has changed. Direct method and grammar translation approaches were common in the first half of the 1900s. Students studied the grammatical system of the new language and translated what they wanted to say or write from their native language into the new one or vice versa. Audiolingual and audiovisual methods surfaced in the middle of the century, and many students learned a new language by listening to tapes of speakers of the language, seeing filmstrips, practicing controlled and meaningless dialogues, and completing grammar or vocabulary substitution drills. In the 1970s and after, the communicative method became more popular: instruction focused on students' using functional language in meaningful ways to promote conversational and survival language skills. For example, students might use the new language to read a bus timetable, to plan a trip, or to seek advice from a peer about course selections for the next school year.

In communicative classrooms, students can discuss material of high interest and topicality, which motivates them to learn and participate in class. Students practice using the new language in realistic scenarios rather than through substitution drills or rote recitals of grammar-driven dialogues. Students are more able to personalize a conversation or a piece of writing. They are explicitly taught to negotiate meaning, ask for clarification, and paraphrase. Students are encouraged to experiment with the language and assume greater responsibility for their learning.

Communicative language teaching was a welcome change for many students because the oral activities, readings, and writings are engaging and purposeful. However, educators realized that the communicative approach did not provide all the skills and vocabulary students needed once they exited an ESL program. So K–12 and postsecondary language educators developed the content-based ESL approach to better prepare ELLs for their transition to subject-area classes (Faltis and Wolfe 1999; Ovando and Collier 1998; Snow and Brinton 1997). Content-based ESL classes are designed for English language learners only and are taught by ESL teachers. Content-based ESL teachers seek to develop the students' English language proficiency by incorporating information from subject areas that students are studying or from courses they may have missed if they are new to the school system. The content-based ESL teachers often use thematic or interdisciplinary units and address key topics found in grade-level subject curricula with a focus on language development.

Content-based ESL instruction alone, however, has not been enough to close the achievement gap. The growth in numbers of students learning English as an additional language, the shortage of qualified ESL and bilingual teachers, the pressure of meeting academic standards, and the focus on high-stakes assessments have generated the need to teach content to these students outside the ESL classrooms. As a result, ESL educators and content teachers developed the sheltered content instruction approach (Crandall 1987; Short 1994). In some regions of the United States, this approach is known as specially designed academic instruction in English (SDAIE).

Through sheltered instruction, ELLs participate in a content course with grade-level objectives, but teachers modify the instruction to make the information more comprehensible. Usually, sheltered-classroom or subject-area teachers—such as Mr. Hafiz—have participated in specialized professional development to learn to teach these courses. Common sheltered instruction techniques include slower speech, use of visuals and demonstrations, connections to student experiences, and use of supplementary materials. In addition, recent research on a specific sheltered approach, the SIOP Model (Echevarria et al. 2004), described in Chapter 7, indicates the value of targeted attention to language objectives in each lesson, explicit instruction in academic reading and writing strategies, and extensive, enriched vocabulary development. Content-based ESL and sheltered instruction courses are proving to be a promising combination when implemented throughout a school.

Ideally, language and content teachers collaborate in schools. Content-based ESL teachers address all the objectives in the district or state ESL or English language development curriculum. They design lessons to target key content area vocabulary, grade-level topics, and the academic tasks students need to learn (e.g., how to follow directions to complete an experiment, how to take notes from reference materials, how to give an oral presentation). ESL teachers must teach language-learning strategies to students as well (e.g., using cognates to determine meanings of unknown words, previewing headings in a chapter before reading).

Sheltered teachers cover the curriculum of their subject area and teach language objectives that apply to the content area lessons. For example, if students are expected to record observations during a science experiment, the sheltered science teacher may need to review descriptive adjectives and sequence terms with the students in advance. Or, if the students have to classify and compare animals, teachers need to make sure students know how to form comparatives or use expressions like "both …and …," "on the one hand …, on the other hand …," and "in contrast…." The sheltered teacher does not replace the ESL teacher but reinforces the use of academic English in ways particular to each subject.

Although current practice and research favor content-based ESL and sheltered instruction program models, preservice teacher education for elementary and secondary teachers outside the ESL or bilingual education field has not kept pace. As noted, most states do not require all teacher candidates to take courses in ESL methods or sheltered instruction techniques. As a result, the professional development burden often falls to the school districts. It is through inservice teacher development or graduate coursework that most teachers learn strategies and techniques for working with ELLs in content courses (National Center for Education Statistics 2002). ESL specialists, in turn, need professional development to learn the content they are teaching in content-based ESL courses.

At the elementary school level, all classroom teachers with ELL students should have training in sheltered instruction. Because this has not yet happened, a relatively new approach at this level is the co-teaching model. In this case, the ESL teacher spends part of the day in the regular classroom, sharing the instructional load with the grade-level teacher, in ways such as being responsible for vocabulary development or pre-reading activities with the whole class and providing additional attention to ESL students during other tasks.

At the secondary school level, ESL students may have one or more periods of ESL (depending on their proficiency level) and also have sheltered science, math, and social studies classes. Ideally, the ESL period offers a content-based curriculum. In a number of schools, students may make the transition out of one sheltered course at a time, as they demonstrate both content and language competence in that area (Short 1999).

Changes in Science Education

Science education, like ESL education, has evolved over the years. Traditionally, teachers taught science to students through lecture and textbook readings, with some laboratory demonstrations by the teacher or carefully controlled experiments by student groups. This approach allowed teachers to convey a large amount of information expediently to students, cover the curriculum in predictable time frames, and accommodate limited supplies of equipment and materials.

But over time, science educators have debated the role of content versus process. Besides imparting facts about scientific concepts and processes, educators have recognized the need for students to learn to think and act like scientists so they can explore the scientific process more fully. Most educators today believe both content and process are essential and that these key elements should not be separated in science teaching.

In the past, and to some extent now, science teaching in many middle and high schools has followed textbook programs featuring cookbook-type laboratory experiences for students. These programs do not ask for much exploration and do not pose essential questions for students to address. Students learn some basic laboratory skills but do little critical thinking or actual planning of experiments. Some schools have moved to discovery learning and guided inquiry, which offers students more choice in such activities as designing experiments, generating testable questions, and evaluating different solutions. In some classes, student-designed inquiry takes place with a great deal of student independence evident in selecting a topic, generating a test question and hypothesis, designing the experiment, interpreting results, drawing conclusions, and reconsidering the experiment.

In the elementary schools, science has been almost nonexistent in the past, except for what might be termed "episodic" science based on teacher interest. In the 1960s, a push for hands-on science curricula took place, and new programs were developed, such as the Science Curriculum Improvement Study (SCIS), Science: A Process Approach (SAPA), and the Elementary Science

Study (ESS). Elementary science educators realized that hands-on experiences play an important role in students' concept development. SCIS introduced the learning cycle as the design for teacher-learner interaction set forth in three overlapping stages: exploration, invention, and discovery (Atkin and Karplus 1962; Karplus 1974). In exploration, the students "first explore materials with minimal guidance, instruction or specific questions." Then, in the invention phase, the teacher helps students to develop "new conceptual structures to interpret the observations." Because few students can clearly state new concepts by themselves, the teacher provides the definitions and terms needed. "Discovery is used to describe activities in which a student finds a new application" and develops his or her real understanding through experience with the concepts the teacher has selected (Delta Education 1998). Unfortunately, in practice, some hands-on experiences in SCIS lessons resulted in students playing around with the materials, not partaking in true scientific learning. What was needed was a minds-on component. An effective environment for learning science would come alive with activity, but also be rooted in the concepts and facts of science as students apply a range of scientific skills to their lives.

The National Science Education Standards (NSES) (NRC 1996), described in Chapter 9, provided a realistic road map for the science education community and helped educators understand the importance of scientific literacy in the 21st century. Among the content was a strong theme that just memorizing facts will not produce people who can use scientific knowledge for informed decision making. The NSES called for a "new order where teachers and students can work together as active learners" (p. 27). The Standards envision the teacher's role as facilitator of student learning through inquiry approaches and student-centered curricula. "Inquiry into authentic questions generated from student experiences, the central strategy for science teaching" (p. 31) is clearly articulated in these science Standards, along with the idea that students need to be provided with opportunities to engage in inquiries with authentic questions.

The extent to which the ideas set forth by the NSES are practiced in our schools today is uncertain, however, and very much dependent upon a school district and its leadership. Science teaching in today's classroom is quite varied. Districts have diverse needs and diverse learners, but must be accountable for the academic performance of all students. A science teacher's dilemma lies in determining what the most effective method of teaching science would be. One recommended best practice that meets the demands of school accountability and incorporates some NSES ideas is guided inquiry, described in Chapters 5 and 7 of this book and further elaborated in Thier and Daviss (2001).

Guided inquiry differs from conventional hands-on science learning because, after students complete an assigned activity, they are encouraged to design their own projects and investigations to continue exploring the topic. Through these self-selected activities, facilitated by a teacher, students pursue questions relevant to them. Working in this way, students link key ideas, reconsider their own theories, and perhaps even satisfy their curiosity in order to achieve a deeper, more enduring level of understanding.

Research has shown that students learn better when they experience something by doing it instead of reading about it in a textbook or hearing about it in a lecture. When students work like scientists, they use language to organize, recognize, and internalize the concepts, principles, and information that they encounter through activities. By providing literacy opportunities for students in science, educators enrich the context for both subjects so students can more effectively expand their personal structures of science knowledge by improving their language skills (Thier and Daviss 2002).

Studies have also shown that true learning takes place only when students engage with information and processes deeply enough to weave that content into their personal views and understandings of how the world works (Harlen 2000). The concept of guided inquiry gives equal weight to knowledge and skills, scientific facts, and processes. It emphasizes concepts more than rote formulas and learning science in a personal and social context rather than through abstractions. To take students beyond the formulaic aspects of science, teachers must rely on students' language skills. By embedding an inquiry within both the context of students' lives and strong science content, then sequencing investigations as part of a larger curricular design, teachers can reach their instructional goals for science and English at the same time.

Teaching Science

In this information age, fueled by technology, students need not only to understand the concepts and processes of science, but they also must be able to apply them to a range of scientific skills to be effective members of the 21st century. To be scientifically literate, students need to possess a set of skills that merges the knowledge of science concepts, facts, and processes with the ability to use language to articulate and communicate about these ideas.

Teachers need to ensure that students internalize scientific habits of mind, such as using evidence to separate opinion from fact. If students are to become adults capable of making informed choices and taking effective action, then they must absorb such habits into their regular patterns of thought so that those habits become an enduring part of their thinking long after their time in school is over.

Given the importance of science to the future, educators have the opportunity to merge the teaching of both science and language literacy to strengthen students' skills. As Scott (1992) writes in *Science and Language Links*, "Language plays [roles] in science learning ... science can be used to develop children's language, and ... increased knowledge of language goes hand in hand with the development of scientific ideas" (p. ix). Researchers have found that students learn science better when they write, or speak, about their thinking, and that the act of writing, or speaking, "may force integration of new ideas and relationships with prior knowledge. This forced integration may also provide feedback to the writer and encourage personal involvement" with what is being studied (Fellows 1994). In the classroom, therefore, science and language become interdependent, in part because each is based on processes and skills that are mir-

rored in the other. These reciprocal skills give teachers and students a unique leverage: by merging science and language in the classroom, teachers can help students learn both more effectively.

It is important, however, for science educators to realize that English language learners are studying new, challenging science curricula in and through a new language. Many students today struggle to meet high academic standards, but ELLs have the added complexity of having to learn, comprehend, and apply the academic English that the teacher and textbooks use to deliver important information. Teachers of ELLs must take into consideration their unique second language acquisition needs and design and deliver lessons that are meaningful and appropriate, as Mr. Hafiz did in the scenario at the beginning of this chapter. Moreover, in order to tailor instruction appropriately, teachers must also recognize that not all English language learners have the same linguistic or educational backgrounds.

English language learners are identified as diverse because they represent different cultures, but they also represent different language backgrounds, levels of native language and English literacy, and educational experiences. Some ELLs are immigrants who arrive in U.S. schools at or above grade-level in terms of their educational backgrounds and literacy skills. They have studied a similar series of science courses in their native country and have the conceptual knowledge needed to succeed in a U.S. science classroom. What they need is to learn the English vocabulary and language structure in order to continue learning and articulate their understanding to teachers and classmates.

Other students have had interrupted schooling in their native countries and may be several years behind their grade-level peers. Basic scientific understandings that textbooks and curricula assume students know when they enter grade 6, for instance, may be lacking. Besides English language development, these students need instruction in the basic content of science that provides a foundation for the complex science to come in later school years. They also need to learn classroom practices of science courses (i.e., following lab directions, organizing equipment, and recording observations). In particular, they need teachers who know how to bridge gaps in their knowledge while moving them forward with the grade-level curriculum.

Students who face the most challenging situation are those who are illiterate or semiliterate in their native language. They may arrive in school beyond the age when "learning to read" is taught. Yet, they need to learn how to read and write in English, and then how to "read to learn" for upper elementary and middle school classes. Usually their illiteracy is a result of very limited or nonexistent schooling opportunities, so they also have major gaps in their academic knowledge that need to be addressed. Although this group of learners is not a majority of ELLs, it is the group that many teachers have had the least preparation for working with.

Although these are generalized profiles, they point out the need for an awareness among teachers that assumptions about underlying language skills are not necessarily valid with ELLs. Although there is increased emphasis in the science education field on promoting more literacy in science, current recommended practices for native English speakers should be adjusted when applied to ELLs. ELLs are likely to gain conversational fluency in English within one to two years of daily exposure in the United States, but the language that is critical for educational success—academic language—is more complex and develops more slowly and systematically in academic settings (Cummins 2000). Research has shown that it may take students from four to ten years of study, depending on the background factors described above, before they are proficient in academic English (Thomas and Collier 2002). Science educators must realize that even if a student has mastered conversational English, that student may still be acquiring proficiency in reading scientific text and communicating in academic discourse, especially given the highly complex terminology used in science texts in upper elementary and secondary school programs.

Lemke (1990) has examined scientific discourse and noted that "scientific language has a preference in its grammar for using the passive voice... people tend to disappear from science as actors or agents . . . [and there is] a grammatical preference for using abstract nouns derived from verbs" (p. 130). He has criticized these stylistic norms, which render science less accessible and less engaging to students. "[T]eachers tend to leave much of the semantics and grammar of scientific language completely implicit" (p. 170). Both Lemke and Gibbons (2003) have argued instead that teachers should converse directly with students about scientific talk, introduce semantic relationships among scientific terms, and give students more practice in speaking about science. They have recommended using informal or colloquial speech initially so students understand the vocabulary and concepts being taught, and later teaching the ELLs the necessary technical terms, grammatical expressions, and discourse patterns, such as use of argumentation.

In a similar manner, Warren and Rosebery (1995) demonstrated the importance of introducing students to the discourse conventions of science, helping them practice not only the habits of mind of scientists but also the discourse norms (e.g., phrasing hypotheses and supporting claims with evidence). Consider, for example, using oral discourse to discuss the results of an experiment before asking students to write a lab report. This is a useful technique for native English speakers, but it must be applied carefully with ELLs, especially those at beginning and intermediate levels of proficiency if they do not have scientific vocabulary or ways of expressing observations and conclusions in their English language repertoire yet.

Teachers also need to keep in mind the types of classroom cultures students have experienced. As Lemke (1990) noted, competence in content classes requires more than mastery of the subject matter topics; it requires an understanding of and facility with the type of texts and procedures for spoken and written interaction and the skills to participate in class activities. Many science classes incorporate inquiry lessons. These engage students in discovering scientific principles and conducting science

experiments in ways similar to scientists. But some ESL students who are recent immigrants may never have experienced an inquiry lesson. They may never have had the opportunity to conduct an experiment by manipulating scientific equipment and materials. They may have learned science through rote memorization of teacher lectures or textbook chapters. Therefore, teachers will need to introduce ELLs to a classroom culture in which students are expected to participate orally, work in cooperative groups, solve problems, conduct experiments, generate hypotheses, and express opinions. Because communication patterns in class may be different from those in students' native cultures, teachers need to be sensitive to and build upon culturally different ways of learning, behaving, and using language.

Promoting Scientific Literacy Among English Language Learners

In this activity, "My Sweet Tooth" (SEPUP 1997), students gather evidence and make decisions about the taste, nutritional value, and health implications of sugar and its substitutes. Ms. Feehan, the fourth-grade teacher, *begins the lesson by showing students packets of sugar and other sweeteners from a restaurant. She introduces the words* sugar, sweet, sweeteners, real, *and* artificial *and asks students to give examples of foods that represent those terms. She records their ideas on the board.*

Window
Into the
Classroom

Then she asks students to brainstorm ideas about sweeteners in their diets. One notes that sugar is a source of energy and it makes foods taste better. Another says, "Sugar makes you fat and gives you cavities, so our family uses artificial sweeteners." A third responds, "Yuck! That stuff's made out of chemicals." Ms. Feehan asks two ESL students what they use to make foods sweet in their home countries. One talks about sugar cane and another describes honey. The conversation helps students begin to define the properties of, and differences among, sweeteners.

Next, the activity leads students through three stages. In each, students use language in slightly different forms to gather information, evaluate it, and make evidence-based decisions. In the first stage, students note physical details; in the second, they record personal impressions; in the third, they read to fill gaps in their knowledge. For each stage, the teacher models the procedures students should use with a different substance—for example, coffee granules or tea leaves.

Working in small groups, students begin by looking closely at small samples of sugar and two artificial sweeteners. Students note the details of the materials' physical appearance, first by looking at them unaided and then by looking at the sweeteners through magnifiers. In discussions with their group members, students compare and contrast the magnified and unmagnified appearances of various substances. They discuss and then list adjectives for each substance in a three-column chart (one column per substance).

In the second stage, students make solutions of each substance, taste each one[1], and discuss and record their personal preferences along with the reasons for their choices in the chart. Ms. Feehan asks students to choose which sweetener they would use at home and explain their reasons to a partner. The discussions help students use spoken language to explore and organize their knowledge and thoughts about sweeteners and to articulate them based on different kinds of evidence. ELLs can discuss their ideas in a less stressful environment—with a peer—rather than aloud to the whole class. In addition, the students use language to identify information they do not have but would need in order to make a better-informed choice.

At this point—when students are motivated to find out more—Ms. Feehan asks whether the students would like additional information before finalizing their decisions and, if so, what kind of information they would like. Students may know which sweetener tastes best to them, but perhaps the discussion has raised concerns, for example, "I like this one best, but is it good for me?"

Ms. Feehan then distributes a page of background information about different sweeteners for students to read. She pairs ESL students with native English speakers to read quietly together. She then asks the students if their choices have changed, and offers to help students conduct additional library or internet or other research on these and other sweeteners.

(excerpted from The New Science Literacy: Using Language Skills to Help Students Learn Science. Thier and Daviss 2002).

When teachers take advantage of ready opportunities to unite science and language to strengthen each other, three benefits result. First, in elementary grades, science gives meaning and purpose to literacy activities by providing a rich field of content that students are naturally curious about such as their bodies, the sky, and animals. When literacy skills are linked to science content, students have personal, practical motivation to master language as a tool that can help them answer their questions about the world around them.

Second, the stronger a student's literacy skills, the stronger the student's grasp of science will be. For those people who are not professional scientists, scientific concepts, principles, and information are most easily expressed and understood in nonmathematical or technical terms. Language becomes the primary avenue that students must travel to arrive at scientific understanding.

Third, as noted earlier, teachers in middle and high school cannot assume that students, especially ELLs, entering their science courses have an adequate vocabulary and the necessary skills to decode print and draw meaning from language. By employing a few, well-chosen literacy techniques and by teaching students some language learning strategies, science teachers can help students improve their reading comprehension and their achievement in science. Teachers who have done so report that the time they invested has been more than repaid in students' accelerated academic progress and in their increased ability to learn independently, without the repeated teacher intervention and monitoring that often accompanies traditional instructional methods (Thier and Daviss 2002).

Just as language clarifies and communicates the meaning of science, science can strengthen the meaning that students find in language studies. Research has shown that the acquisition of literacy skills is significantly enhanced when those skills are used for specific purposes within a meaningful and stimulating context. Language instruction in concert with materials-centered science activities can provide just that purposeful

environment needed to reinforce students' emerging literacy skills. For ELLs this is particularly noteworthy because the hands-on nature of many science activities provides context for the information being conveyed that is less language dependent than reading a textbook or listening to a teacher lecture would be. Studies have also shown that student learning in science is improved by the introduction of literacy-related activities such as process writing (Bredderman 1983; Fellows 1994; Holliday, Yore, and Alverman 1994; Rowe 1996). As Holliday et al. (1994) note, "Hands-on experiences are necessary, but not sufficient, to learn many counterintuitive science concepts. Likewise, language is necessary, but not sufficient, in initial learning of abstract concepts. The important factors are: . . . what types of thinking and strategies are mutually beneficial in reading, writing, and science?" (p. 877).

A resource for teaching science and literacy skills may be found in the New Standards project, a joint venture of the Learning Research and Development Center at the University of Pittsburgh and the National Center for Education and the Economy (2000). These generic skills, modified specifically for use with inquiry-based science programs, are commonly used as frameworks or guidelines by school districts and materials developers. With some additional support, they may be applied in classes with English language learners.

Throughout the "My Sweet Tooth" activity described in the second "Window Into the Classroom," students use many of the reciprocal processes that literacy and science share. Looking at the sweeteners, they note physical details and then compare and contrast the substances' appearances. They gain experience in understanding language operationally—experiencing a concept before learning its abstract name and definition, thereby being able to viscerally associate a term with a concrete meaning. They infer, from scientific studies using animals, how artificial sweeteners might affect humans. They draw evidence-based conclusions about which sweeteners they personally would or would not use. Through their observations and discussions, they use language to sort through evidence and to distinguish facts from opinion. As they read, write, discuss and debate, listen critically to other students, and work through a rudimentary scientific investigation, they also begin to understand that evidence can be a powerful factor in understanding the world.

In addition, the activity's links between science and language can be broadened and strengthened through additional, student-designed investigations. For example, students may conduct a classwide taste test between a regular cola and its low-calorie alternative, then graph the number of students who preferred the taste of each (or found no difference between them). Students also could write reports about the origins, benefits and drawbacks, and health implications of corn syrup, honey, cane sugar, and other sweeteners.

As part of "My Sweet Tooth," teachers often encourage students to search the library or the internet for more information about sweeteners. Usually, students find an array of articles, some with conflicting messages. To interpret the materials' collective meanings as accurately as possible, the students must evaluate the information and presentation strategies of the authors. Was a study declar-

ing a sweetener to be safe for humans paid for or written by the company that makes it? Does the study include negative information about the sweetener's effects on people that a previous report found? ELLs may need additional assistance in understanding subtle differences among the language used in commercial claims. This kind of analysis joins conventional reading comprehension—understanding the meaning expressed through the words—to the comprehension of an author's intention and purpose. Through teacher scaffolding, ELLs and other students can learn to read between the lines and be able to consume information more critically.

To provide additional language practice through this activity, students can be encouraged to write in a journal, thus linking literacy and science. This would serve as a record of the evidence they gather and of their questions and speculations about what the evidence means and how ideas relate to one another. A journal helps a student create a personal scaffold on which to assemble data, observations, other evidence, reflections, and questions awaiting additional data for answers. Students use and experiment with language in a journal to knit together their scientific data and evidence, their ideas, and their inferences and conclusions. In their journals, students can record answers to questions like "What new insights do I have that I didn't have before? What do these new insights mean for 'My Sweet Tooth' and for me when I select the sweetener I will use? What don't I know, and how will I find out?"

In sum, by making science a key element in strengthening language and literacy skills, teachers can demonstrate that emphasizing science—not ignoring or de-emphasizing it—is a crucial step in raising students' achievement scores on standardized language exams, currently a dominant aim among thousands of school districts. By reinforcing language and literacy skills within a meaningful context, science can retain its place in a curriculum that, due to external pressures, too often skews its emphasis toward other disciplines.

Research-Based Practices and Programs

With the focus on school reform over the past decade, many research studies have looked at effective practices for improving the educational achievement of at-risk students, including English language learners. The federal government, through the U.S. Department of Education and the National Science Foundation, has played a critical role in funding this type of research. The following are promising interventions.

Center for Research on Education, Diversity and Excellence Studies

The National Center for Research on Education, Diversity and Excellence (CREDE at *www.crede.org*) was charged with studying ways to improve the education of students at risk for success in school due to language, cultural, or geographic diversity. Funded by the U.S. Department of Education, Office of Educational Research and Improvement (now the Institute for Education Sciences), this center engaged in seven years of research focused on program designs, instructional practices, professional development models, family and community involvement, and school reform models.

Two complementary CREDE studies specifically examined pedagogical principles for teaching content to ELLs in meaningful ways that address their cultural and linguistic diversity. The first study led to the development of the Sheltered Instruction Observation Protocol (SIOP) Model, which has been applied across core content areas. The second, known as the Five Standards for Pedagogy, likewise has application across the curriculum.

The SIOP Model

The SIOP Model, which has been described in detail in Chapter 8, is a research-based model of sheltered instruction developed by researchers at the Center for Applied Linguistics and the California State University, Long Beach for CREDE (Echevarria et al. 2004). Field-tested over seven years in collaboration with middle school teachers, the SIOP Model incorporates best practices for teaching academic English and provides teachers with a coherent, usable approach for improving the achievement of their students. The CREDE research on the SIOP Model showed that ELLs whose teachers were trained in implementing the SIOP Model performed statistically significantly better on an academic writing assessment than a comparison group of ELLs whose teachers had had no exposure to the model (Echevarria et al. forthcoming). A separate study found that the Sheltered Instruction Observation Protocol instrument, which was developed to measure teacher fidelity while implementing the SIOP Model, was a highly reliable and valid measure of sheltered instruction (Guarino et al. 2001).

The SIOP Model includes features recommended for high-quality instruction for all students, such as cooperative learning, reading comprehension strategies, and differentiated instruction. However, the SIOP Model adds key techniques for the academic success of ELLs, such as including language objectives in every content lesson, developing background knowledge, highlighting content-related vocabulary, and emphasizing academic literacy practice. The model offers a flexible framework for organizing instruction yet provides teachers with specific lesson features that, when implemented consistently and to a high degree, lead to improved academic outcomes for ELLs. (See *www.cal.org/siop* and *www.siopinstitute.net* for more information.)

Five Standards of Pedagogy

Research has shown that the Five Standards of Pedagogy (Tharp 1997; Tharp et al. 2003) have led to positive learning outcomes for at-risk students. These standards are generic principles applicable across the curriculum when culturally and linguistically diverse students are present. They include

1. facilitate learning through joint productive activity among teachers and students,

2. develop students' competence in the language and literacy of instruction throughout all instructional activities,

3. contextualize teaching and curriculum in the experiences and skills of home and community,

4. challenge students toward cognitive complexity, and

5. engage students through dialogue, especially the instructional conversation.

Through multiyear partnerships with demonstration schools, CREDE researchers and coaches have worked with teachers to guide their implementation of the five principles into classroom instruction. Over time, as teachers become skilled in the instructional practices, they begin to structure lessons around multiple, simultaneous activity centers. CREDE researchers developed an observation tool, the Standards Performance Continuum (SPC), to measure level and fidelity of implementation and gathered student achievement data to determine outcomes. The findings indicated that students in classes with teachers who scored high on the SPC had significantly better achievement gains than students in classes with teachers who did not score as well. (See *www.crede.org* for more information.)

National Science Foundation Grant Projects

Over the past 10 years, a small but growing number of studies funded by the National Science Foundation have examined how to teach science to diverse learners; some have focused on English language learners. What follows are three project descriptions that illustrate the research directions that have been taken in recent years. (See individual granting descriptions at *www.NSF.org*)

California: The Valle Imperial Project in Science (VIPS)

This project, focused on inquiry-based science, has developed a successful track record with students in Southern California, most of whom are Hispanic. The VIPS project incorporates hands-on, research-based elementary science curricula, drawn in part from the

National Science Resource Center's Science and Technology for Children program, sustained professional development for teachers (a minimum of 100 hours), and in-class support from resource specialists while teachers implement the curriculum. Curriculum and materials included science kits and student notebooks. Fourteen districts in the Imperial Valley region of California have joined with San Diego State University to carry out this effort. Results on the science section of the Stanford Achievement Test, ninth edition, for fourth and sixth graders, and on the sixth-grade district writing proficiency test, showed that students in the VIPS program outperformed nonparticipating students. (More information can be found at *www.vipscience. com* and *www.carolina.com/stc/publications/ evidence/vips.pdf*)

Florida: Promoting Science Among English Language Learners (P-SELL)

A series of National Science Foundation grants have funded research on science learning among culturally and linguistically diverse students in a large urban school district in southern Florida over the past decade. These research studies have sought to promote science learning and English language and literacy development for elementary students of African American, Asian American, Haitian American, Hispanic, and White Non-Hispanic backgrounds through teacher development and materials development. The studies have involved teachers and students in grades three through five at a number of elementary schools. The research examines the interplay of three domains: (a) science knowledge and inquiry practices, (b) English language and literacy development,

and (c) students' home language and culture (Lee 2004; Lee and Fradd 1998). Over time, instructional interventions that include lesson plans and materials have been developed, along with workshops for teachers that reflect effective science classroom practice. The results show positive outcomes in terms of teacher change (i.e., increased use of hands-on and inquiry-based science instruction) and student achievement (Cuevas et al. forthcoming). The emphasis is on integrating literacy into science instruction for ELLs and other students with limited literacy development and science experience, but teachers experienced difficulty in relating science to students' linguistic and cultural experiences (Luykx et al. 2004).

Arizona: Communication in Science Inquiry Project (CISIP)

The Communication in Science Inquiry Project (CISIP) is a five-year project aimed at creating and disseminating professional development materials for science and English faculty at middle schools, high schools, and community colleges in the Phoenix area of Arizona. The goal is to create professional development materials that show science and English teachers how science inquiry lessons can integrate scientific writing, writing-to-learn processes, and science language acquisition techniques into instruction. The science content includes physics, chemistry, geology/Earth and space sciences, and life sciences. It is expected that teachers and students who use these materials increase their understanding of scientific concepts and their capacity to write scientifically with greater fluency and complexity. The project places particular emphasis on helping students who are English language learners and their teachers. The project involves a collaboration of faculty from Maricopa Community College, Arizona State University, and others. The team will work with public school districts to develop, pilot, field test, research, and evaluate the materials. The formative evaluation will help refine the materials. At the end of the project, which began in 2003, the team plans to have at least 24 field-tested lesson models in an integrated curriculum model that includes writing-to-learn processes and scientific writing illustrations. Part of the professional development materials will address language acquisition.

These ongoing projects place an emphasis on teacher training, resource support for teachers, partnering between schools and universities, collaboration between teachers, materials development, and systematic evaluation. The results of these projects and many other programs in the United States and abroad suggest that all students benefit when language and science instruction are integrated.

Conclusion

Helping English language learners succeed in science classes is important for their overall academic achievement in school and beyond. We want students to become informed citizens, capable of making scientifically literate decisions about their lives in the future. We also want to take advantage of the opportunity that combining language and literacy learning with science offers. By merging language development, literacy, and science consistently, teachers can demonstrate that their subject has educational power across the curriculum.

This book is a collaborative effort between science and language educators. Integrating strengths from both fields gives teachers the opportunity to better understand the needs of diverse students and to integrate science and language teaching practices. The ideas incorporated in this book will, we hope, bring the worlds of science and language instruction together and provide new opportunities for teachers to collaborate and work together for this most important responsibility. Meeting the challenges of ensuring academic success for linguistically and culturally diverse students in our classrooms can be an exciting learning opportunity, as well as a challenging experience for all.

The reality is that collaboration and professional development are limited by the many educational constraints facing schools and teachers daily. The school day is packed full of required lessons and activities, and at times, test preparation exercises. School budgets and other resources, especially teachers' time and energy, are already stretched to the limit. It is important for legislators, administrators, and educators at every level to take action. Time, training, and resources must be provided for teachers if we agree with the NSTA (2000) position statement that "ALL children can learn and be successful in science and our nation must cultivate and harvest the minds of ALL children and provide the resources to do so."

Endnote

[1]Teachers need to check if any students have PKU, a genetic condition that makes them sensitive to aspartame, before this stage of the lesson. Such students should not taste substances containing this artificial sweetener.

References

American Association for the Advancement of Science. 1967. *Science—A process approach.* New York: Xerox Division, Ginn

Atkin, J., and R. Karplus.1962. Discovery or invention? *The Science Teacher* 29 (5): 45.

Bredderman, T. 1983. Effects of activity-based elementary science on student outcomes: A quantitative synthesis. *Review of Educational Research* 53 (4).

California Department of Education. 2004. Statewide Stanford 9 test results for reading: Number of students tested and percent scoring at or above the 50th percentile ranking. Retrieved February 23, 2004, from *www.cde.ca.gov/dataquest/* (no longer available).

Crandall, J.A., ed. 1987. *ESL in content-area instruction.* Englewood Cliffs, NJ: Prentice Hall Regents.

Cuevas, P., O. Lee, J. Hart, and R. Deaktor. Forthcoming. Improving science inquiry with elementary students of diverse backgrounds. *Journal of Research in Science Teaching.*

Cummins, J. 2000. *Language, power and pedagogy: Bilingual children in the crossfire.* Clevedon, England: Multilingual Matters.

Delta Education. 1998. *The learning cycle: Guided inquiry.* Hudson, NH: Author.

Echevarria, J., D. Short, and K. Powers. Forthcoming. School reform and standards-based education: An instructional model for English language learners. *Journal of Educational Research.*

Echevarria, J., M. Vogt, and D. Short. 2004. *Making content comprehensible to English learners: The SIOP model,* 2nd ed. Boston: Pearson/Allyn and Bacon.

Educational Development Center. 1969. *Elementary science study.* Manchester, MO: Webster Division, McGraw-Hill.

Faltis, C., and P. Wolfe, eds. 1999. *So much to say: Adolescents, bilingualism and ESL in the secondary school.* New York: Teachers College Press.

Fellows, N. J. 1994. A window into thinking: Using student writing to understand conceptual change in science learning. *Journal of Research in Science Teaching* 31.

Gibbons, P. 2003. Mediating language learning: Teacher interactions with ESL students in a content-based classroom. *TESOL Quarterly* 37 (2): 247–273.

Guarino, A, J. Echevarria, D. Short, J. Schick, S. Forbes, and R. Rueda. 2001. The sheltered instruction observation protocol. *Journal of Research in Education* 11(1): 138–140.

Harlen, W. 2000. *Building for conceptual understanding in science.* Berkeley, CA: Lawrence Hall of Science, University of California.

Holliday, W., L. Yore, and D. Alverman. 1994. The reading-science learning-writing connection: Breakthroughs, barriers, and promises." *Journal of Research in Science Teaching* 31: 877–894.

Karplus, R. 1974. The learning cycle. In *The SCIS Teachers Handbook.* Berkeley, CA: Regents of the University of California.

Latinos in education: Early childhood, elementary, undergraduate, graduate. 1999. Washington, DC: White House Initiative on Educational Excellence for Hispanic Americans.

Learning Research and Development Center, and the National Center on Education and the Economy. 2000. *New standards: Performance standards and assessments for the schools.* Pittsburgh, PA: Learning Research and Development Center, University of Pittsburgh.

Lee, O. 2004. Teacher change in beliefs and practices in science and literacy instruction with English language learners. *Journal of Research in Science Teaching* 41(1): 65–93.

Lee, O., and S. Fradd. 1998. Science for all, including students from non-English language backgrounds. *Educational Researcher* 27(4): 12–21.

Lemke, J. 1990. *Talking science: Language, learning and*

values. New York: Ablex.

Luykx, A., P. Cuevas, J. Lambert, and O. Lee. 2004. Unpacking teachers' "resistance" to integrating students' language and culture into elementary science instruction. In *Preparing prospective mathematics and science teachers to teach for diversity: Promising strategies for transformative action,* eds. A. Rodríguez and R. S. Kitchen, 119–141. Mahwah, NJ: Erlbaum.

National Center for Education Statistics. 2002. *Schools and staffing survey, 1999–2000: Overview of the data for public, private, public charter, and Bureau of Indian Affairs elementary and secondary schools.* (NCES 2002-313). Washington, DC: U.S. Department of Education, National Center for Education Statistics.

National Commission on Teaching and America's Future (NCTAF). 1996. *What matters most: Teaching for America's future.* New York: Teachers College Press.

National Research Council. 1996. *National Science Education Standards.* Washington, DC: National Academy Press.

National Science Teachers Association. 2000. Position statement on multiculturalism. Available online at *www.nsta.org/positionstatement&psid=21.*

National Training Laboratories. n.d. The learning pyramid. In *Why use active learning?* 2000. Abilene, TX: The Abilene Christian University. Posted at *www.acu.edu/cte/activelearning/whyuseal2*.htm.

Ovando, C., and V. Collier. 1998. *Bilingual and ESL classrooms,* 2nd. ed. Boston: McGraw-Hill.

Saul, E. W., ed. 2004. *Crossing borders in literacy and science instruction.* Newark, DE: International Reading Association/National Science Teachers Association.

Science Education for Public Understanding Program (SEPUP). 1997. *Chemicals, health, environment, and me* (CHEM 2). Ronkonkoma, NY: Lab Aids.

Scott, J., ed. 1992. *Science and language links: Classroom*

implications. Portsmouth, NH: Heinemann.

Short, D. 1999. Integrating language and content for effective sheltered instruction programs. In *So much to say: Adolescents, bilingualism and ESL in the secondary school,* eds. C. Faltis and P. Wolfe, 105–137. New York: Teachers College Press.

Short, D. 1994. Expanding middle school horizons: Integrating language, culture and social studies. *TESOL Quarterly* 28 (3): 581–608.

Siegel, H. 2002. Multiculturalism, universalism, and science education: In search of common ground. *Science Education* 86: 803–820.

Snow, C., and G. Biancarosa. 2004. *Reading next: Adolescent literacy development among English language learners.* New York: Alliance for Education and The Carnegie Corporation of New York.

Snow, M. A., and D. Brinton. 1997. *The content-based classroom.* White Plains, NY: Longman.

Teachers of English to Speakers of Other Languages. 1997. *ESL standards for pre-K–12 students.* Alexandria, VA: Author.

Tharp, R. 1997. *From at-risk to excellence: Research, theory, and principles for practice.* Research report No. 1. Santa Cruz, CA, and Washington, DC: CREDE

Tharp, R., P. Estrada, S. Dalton, and L. Yamauchi. 2003. *Teaching transformed: Achieving excellence, fairness, inclusion, and harmony.* Boulder, CO: Westview Press.

Thier, H., and B. Daviss. 2001. *Developing inquiry-based science materials: A guide for educators.* New York: Teachers College Press.

Thier, M., and B. Daviss. 2002. *The new science literacy: Using language skills to help students learn science.* Portsmouth, NH: Heinemann.

Thomas, W., and V. Collier. 2002. *A national study of school effectiveness for language minority students' long-term academic achievement.* Santa Cruz, CA, and Washington, DC: CREDE.

Warren, B., and A. Rosebery. 1995. *This question is just too, too easy: Perspectives from the classroom on accountability in science.* Santa Cruz, CA, and Washington, DC: National Center for Research on Cultural Diversity and Second Language Learning.

Further Reading

Short, D., and J. Echevarria. 2004. Promoting academic literacy for English language learners. *Educational Leadership* 62 (4): 8–13. This article, geared toward administrators, discusses current school reform efforts that focus on the development of strong literacy skills and academic knowledge among the nation's students. All students, including those learning English as an additional language, are called upon to take tests and often have grade-level promotion and high school graduation decisions tied to their scores. At present, many English language learners receive much of their instruction from content-area teachers who have not had sufficient preparation to address second language development needs or to make content instruction comprehensible.

Short, D., J. Hudec, and J. Echevarria. 2002. *Using the SIOP Model: Professional development for sheltered instruction.* Washington, DC: Center for Applied Linguistics. This manual is designed for teacher educators and professional developers who wish to teach others to use the SIOP Model of sheltered instruction in their classrooms. It is a companion piece to the video *The SIOP Model: Sheltered Instruction for Academic Achievement.* In 11 sections, the manual presents strategies, tools, and activities that staff developers can use to instruct teachers in the SIOP components. It also provides a number of black line masters for presentations and workshops.

Thier, M., and B. Daviss. 2001. *Developing inquiry-based science materials: A guide for educators.* New York: Teachers College Press. With a focus throughout on the importance of the teacher as an instructional leader, this book combines instructional materials development and the curriculum in a way that makes the curriculum come alive for teachers and other leaders in science. *Curriculum"* is redefined as the human interaction among teachers and students to accomplish clearly defined goals for instruction and learning.

Thier, M., and B. Daviss. 2002. *The new science literacy: Using language skills to help students learn science.* Portsmouth, NH: Heinemann. This book is a powerful tool for teaching science through language literacy for new or veteran teachers. It presents an encompassing definition of literacy that includes reading, writing, listening, speaking, and critical analysis of media. This book explains how instructional synergy and power result from combining the two subjects and shows how teachers can use practical classroom techniques for integrating these subjects at different grade levels, from elementary to high school. By combining science and language with guided inquiry, teachers can empower students to think and express themselves about science more effectively, improving both their learning and retention.

Additional Media Resources

Hudec, J., and D. Short. 2002a. *Helping English learners succeed: The SIOP model.* Video. Washington, DC: Center for Applied Linguistics. This video provides an introduction to a research-based instructional model that has proven effective with English language learners who are studying content topics while learning English. It is designed for administrators, policymakers, and teachers who would like an introduction to sheltered content instruction and for teacher educators of methodology classes. It is organized around eight components essential for making content comprehensible for English language learners. These components—preparation, building background, comprehensible input, strategies, interaction, practice/application, lesson delivery, and review/assessment—are illustrated in the classroom scenes and elaborated through interviews by the teachers and researchers.

Hudec, J., and D. Short. 2002b. *The SIOP model: Sheltered instruction for academic achievement.* Video. Washington, DC: Center for Applied Linguistics. This new professional development video illustrates the research-based SIOP Model of sheltered instruction, which has raised achievement levels of English language learners. The video is a valuable resource for teacher educators and staff developers to provide ongoing development to teachers of content subjects in strategies and techniques for adapting instruction to the academic needs of these students. Filmed in a documentary style, this video presents each of the SIOP's eight components through footage from the math, science, and social studies classrooms of six exemplary teachers at the elementary, middle, and high school levels. The teachers elaborate on their lessons and SIOP techniques, and teacher educators and researchers describe the theory and research that support the model.

Appendix A

Web References
for National Organizations
and Resources

This section provides electronic links to professional organizations in both language and science education. Many of the websites listed have valuable information and/or links to other resources for educators at all levels. Although only major organizations are listed, individual states may have chapters or branches.

Language Professional Organizations and Resources

Teachers of English to Speakers of Other Languages
www.tesol.org/s_tesol/index.asp

National Association for Bilingual Education
www.nabe.org

American Association of Applied Linguistics (AAAL)
www.aaal.org

International Reading Association (IRA)
www.reading.org

National Council of Teachers of English (NCTE)
www.ncte.org

Center for Applied Linguistics
www.cal.org

National Clearinghouse for English Language Acquisition
www.ncela.gwu.edu

Appendix A

U.S. Department of Education's Office of English Language Acquisition Enhancement, and Academic Achievement for Limited English Proficient Students (OELA)
www.ed.gov/about/offices/list/oela/index.html

Online Directory of ESL Resources
www.cal.org/resources/update.html

Center for Research on Education, Diversity & Excellence Publications and Products
www.cal.org/crede/pubs

CREDE's Five Standards of Effective Pedagogy
www.crede.org/standards/standards.html

Sheltered Instruction Observation Protocol Web Site
www.siopinstitute.net

Center for Equity and Excellence in Education Test Database
http://ceee.gwu.edu/standards_assessments/EAC/HOME.HTM

National Center for Research on Evaluation and Testing (CRESST)
www.cresst.org/index5.htm

National Literacy Panel on Language Minority Children and Youth (NLP)
www.ed.gov/offices/OERI/AtRisk/nlp.html

Science Education Professional Organizations and Resources

Association of Presidential Awardees in Science Teaching (APAST)
http://apast.enc.org/index2.htm

Association of Science Materials Centers (ASMC)
www.kitsupport.org

Association for Multicultural Science Education (AMSE)
http://amse.edhost.org

Association for Science Teacher Education (ASTE) Formerly, The Association for the Education of Teachers of Science (AETS):
http://theaste.org

Council for Elementary Science Education International (CESI)
http://unr.edu/homepage/crowther/cesi.html and *http://CESIscience.org*

Council of State Science Supervisors (CSSS)
http://csss.enc.org

National Association for Research in Science Teaching (NARST)
www.educ.sfu.ca:16080/narstsite

National Earth Science Teachers Association (NESTA)
www.nestanet.org

National Marine Educators Association (NMEA)
www.marine-ed.org

National Middle Level Science Teachers' Association (NMLSTA)
www.nmlsta.org

National Science Education Leadership Association (NSELA)
www.nsela.org

National Science Teachers Association (NSTA)
www.nsta.org

School Science and Mathematics Association (SSMA)
www.ssma.org

Science Education for Students with Disabilities (SESD)
www.sesd.info

Society for College Science Teachers (SCST)
www.scst.suu.edu

Appendix B

Glossary of Science and Language Terms

academic language: Language used in formal school contexts in the learning of academic subject matter.

affective filter: A learner's attitudes that affect the relative success of second language acquisition. Negative feelings such as lack of motivation, lack of self-confidence, and learning anxiety act as filters that hinder and obstruct language learning.

authentic assessment: Multiple forms of assessment that reflect student learning, achievement, and attitudes on instructionally relevant classroom activities.

bilingual education: An educational program in which two languages are used to provide content matter instruction. The proportion of instruction delivered in each language varies between programs depending upon goals.

bilingualism: Bilingualism is the ability to use two languages; however, there may be variation in proficiency in listening, speaking, reading, and writing between the two languages.

cognitive academic language learning approach (CALLA): An instructional model for language and content learning that emphasizes student development of cognitive, metacognitive, and socioaffective learning strategies.

cognitive/academic language proficiency (CALP): The language ability required for academic achievement in context-reduced environments such as classroom lectures and textbook readings (Cummins 1984).

computer-assisted language learning (CALL): Situations where computers and technology accompany learning for English language learners.

constructivism: Steeped in Piagetian theory, this philosophy is a child-centered view of learning where the child constructs new knowledge based upon the negotiation of prior experience and knowledge constructed from experience.

content-based instruction: An instructional approach where content topics are used to develop second language skills.

Appendix B

demonstrations: A way of teaching science where the teacher demonstrates a concept or experiment. Usually used in situations of direct instruction (expository) where safety or cost are issues.

discovery teaching: Similar to inquiry, the students are led through a series of investigations to discover the content. The objective is stated as a conclusion to the lesson.

English as a foreign language (EFL): Situations where English is taught to persons living in countries where English is not the primary language.

English as a second language (ESL): Instruction for English language learners that typically focuses on language and does not involve use of native language.

English as a second language (ESL) standards: Standards that acknowledge the central role of language in the achievement of content and highlight the learning styles and particular instructional and assessment needs of learners who are still developing proficiency in English.

English language development (ELD): Instruction designed for English language learners to develop their listening, speaking, reading, and writing skills in English.

English Language Learners (ELLs, also ELs): Students whose first language is not English and who are in the process of learning English.

English language proficiency standards: Standards that delineate the relationship in the simultaneous acquisition of language and content for English language learners. English language proficiency standards for science represent the interaction between language and content through performance indicators.

exit criteria: A set of criteria for ending special instruction for English language learners and placing them in mainstream, English-only classes.

expository teaching: Teacher as lecturer or demonstrator imparting knowledge to children. Similar to direct instruction.

guided inquiry: In the continuum of inquiry, guided inquiry is most structured. In this model, the teacher leads students through a specific investigation.

home language: The language a student speaks at home with family.

immersion: An approach to teaching language in which the target language is used exclusively for all instruction.

informal education: Education arenas in which learning takes place outside of the walls of the classroom. May include museums, learning centers, and planetariums in formal settings or teaching science outdoors or in other settings where science takes place outside of the classroom.

inquiry teaching: A student-centered approach to teaching in which children are cognitively and kinesthetically engaged with the learning and materials in order to make discoveries of content.

language minority: A person or group that is not from the dominant language group. A language-minority child may be an English learner or be bilingual.

language proficiency: The level of competence that one has in listening, speaking, reading, and writing a language for basic communication as well as academic purposes.

learning cycle: An inquiry-based teaching model centered on students having relevant concrete experiences with a concept before being formally introduced to the idea and then having the opportunity to apply their learning in a new context.

limited English proficient (LEP): The term used by the government and some school districts to identify those students who have insufficient English to succeed in English-only classrooms. The term ELL or EL is frequently used in its place.

mainstream: Classes designed for native or fluent speakers of English, in which no accommodations are made for ELLs.

metalinguistic skills: The ability to talk about language, analyze it, and reflect on it.

National Science Education Standards (NSES): Standards in science education that deal with six domains including: Science Teaching Standards, Standards for Professional Development, Assessment in Science Education, Science Content Standards, Science Education Program Standards, and Science Education System Standards.

native-language support: The use of a child's native language to translate unfamiliar terms or otherwise clarify lessons taught in English.

objectives: The main idea of a lesson stated in a concise statement for student learning.

performance indicators: Acitivities that students may perform to show their progress in meeting a standard.

performance standards: Standards that refer to how well students are meeting content standards that specify how students demonstrate their knowledge and show student progress in meeting standards.

pull-out ESL: A program in which English language learners are "pulled out" of regular, mainstream classrooms for special instruction in English as a second language.

realia: Real objects used in a lesson to help ELLs contextually construct meaning.

rubrics: Descriptive indicators of performance or scoring criteria on a continuum.

scaffolded inquiry: a specific form of the continuum of inquiry ranging from direct inquiry to open-ended inquiry.

scaffolding: A strategy used to provide contextual support for meaning during instruction, such as the use of visuals, graphic organizers, modeling, or questioning. These supports can be withdrawn as learners are able to demonstrate competencies.

science process skills: The skills that are necessary in order to do science, e.g., observing, communicating, classifying, questioning, hypothesizing, etc.

sociocultural context: The interaction of social and cultural elements in learning.

sheltered English: An approach to teaching that makes academic instruction in English understandable to English language learners to help them acquire proficiency in English while at the same time achieving in content areas. In the sheltered classroom, teachers scaffold instruction with the goal of helping students meet grade-level content standards while improving English proficiency.

Sheltered Instruction Observation Protocol (SIOP): The SIOP Model provides a framework containing features for teachers to present standards-based content concepts to ELLs in ways they can understand.

specially designed academic instruction in English (SDAIE): A term for sheltered instruction, SDAIE is a program of instruction in a subject area, delivered in English. Strategies are used that provide English language learners with access to the curriculum.

submersion: The teaching of minority language students through the medium of a majority language without special language assistance.

target language: The language that a child is learning as a second language. The target language for English language learners is English.

total physical response (TPR): A language-learning approach that emphasizes the use of physical activity to provide meaningful opportunities for learning through the use of instructions or commands by the teacher.

Appendix C

Safety for the Science Classroom

John Summers

Teach and Document

Mr. Jackson uses a range of assessments to be sure his students understand, because an eagerly yes-nodding young head may mask a confused mind. To help his students, he instructs using a multitude of techniques.

Window *Into the* Classroom

Mr. Jackson is familiar with ELL teaching methods. He decided that safety could be taught with this style of teaching. He knows he must make each student aware of each day's safety issues. He uses multiple presentation techniques coupled with multiple assessments to ensure that his students understood the importance of safety in the science classroom. Even the newest student, who might have very limited English proficiency, can understand the safety lesson. Mr. Jackson also knows his safety lessons can help ELLs navigate outside the classroom where the electrical outlets, appliances, and other equipment may be unfamiliar.

As a teacher, you know that students come in every possible combination of traits and abilities. You know also that each student is important and that his or her safety is your legal responsibility. Safety is one of the most important things you teach.

Being learners, students test limits. This testing quality of young people is especially important to safety lessons and considerations. Safety lessons must be internalized by each student—become something acted upon in a responsible way. Using all your skills, you must ensure that those lessons are understood, comprehended, and become a way of students' doing things.

If an investigation is risky, you may have to find an alternative. Should a student or a group of students be unwilling or unable to internalize safety considerations, you may have to make an accommodation that ensures their safety as well as that of the rest of the class.

Teaching safer science is much like keeping track of corks bobbing in water. Kids are here one day, out the next. To keep track of who received safety instruction and who did not, prepare a spreadsheet of students and safety instructions. Keep a dated record showing the day each student has received each instruction.

Also document the method by which you ensured a student comprehended the safety instruction. How the lesson is delivered will differ with students at different stages of English proficiency. For any absent student, you must repeat the safety instruction before he or she has access to its corresponding activity.

Appendix C

Liability is real, and you must consider it. Think of the worst case, an accident, and imagine the lawyer for the injured party asking if you have evidence to support your claim that you prepared the student properly and when. For ELL students another category could be important—for you and the student: how did you ensure comprehension?

Your Teaching Area

Mr. Giesen was in his classroom early. While he was thinking through his lesson delivery and the safety practices his students—at varying levels of English proficiency—needed to learn, he also thought about the safety of his teaching area.

He had recently learned of regulations that affected the spaces in which science activities take place. Were the building and room accessible to all? Did the room have two unblocked exits? Was there adequate sanitation? Approved eye protection?

Deciding to invest some more time, Mr. Giesen Googled some pertinent names and ideas and carefully selected government and other reputable websites.

In addition to a myriad other requirements, you must also address federal, state, and county agency regulations that apply to the classroom. Your county health department may already have inspected your science area for adequate ventilation and chemical storage practices. Make sure you are acquainted with these and other regulations: they apply to the classroom teacher and to the school.

Some important requirements have been addressed at the national level. Two of the agencies responsible are NIOSH (National Institute for Occupational Safety and Health) and OSHA (Occupational Safety and Health Administration). OSHA regulates and enforces; NIOSH educates and does safety research.

IDEA (Individuals with Disabilities Act) and ADA (Americans with Disabilities Act) are two government acts that affect classroom safety. IDEA applies to students performing below their abilities, and ADA ensures access to facilities for all members of a community. The acts, which are especially significant in the configuration of classrooms and the science laboratory, apply to every classroom.

Surveying Your Classroom

At end of the day Ms. Antal sat at her desk thinking about safety. Was there something about her classroom or the building she should be aware of? As a middle school teacher, was she taking the right precautions with her small, carefully selected supply of chemicals?

Looking at her classroom she noted a large plant too close to the door. It could be an obstacle. Starting a list, she wrote next to Item 1. "Move the plant away from the back door."

She began to survey her room, concentrating on possible problems. The fruit jars of chemical fertilizer and the dropper bottles of hydrochloric acid, open on the counter, should be locked up. That brought up storage questions. Perhaps she needed a lockable cabinet rather than a too-accessible shelf in a storage closet. Item 2. "Request a lockable cabinet or closet for chemicals."

What had started as a moment of reflection had become a project. Ms. Antal added to her list:

- *Item 3. Check the room for ventilation.*
- *Item 4. Check the room for electrical safety.*
- *Item 5. Involve the administration in determining and meeting safety requirements.*
- *Item 6. Make safety a part of every lesson plan.*

Access to stored chemicals can be a problem especially in multiuse classrooms. Unsecured chemicals and supplies are an invitation to personal, building, and district liability.

You may find other problems. A stuffy classroom environment is a clue to inadequate ventilation. Chemical storage cabinets, depending on their contents, need to be vented to the outside. Classrooms need a constant supply of fresh air, changed according to regulations.

As you survey your classroom, prep areas, and office space, look at your electrical outlets. Are they the grounding, three-prong type? Is the room equipped with ground-fault receptacles, especially around sinks and grounded equipment?

There is more. Science classrooms may need other fixtures such as eyewash and safety showers, both of which must be accessible to anyone. Could a wheelchair-bound student, an ELL student with limited comprehension, or someone on crutches, or anyone else in the room reach these safety devices quickly?

Working with the custodians is another critical part of safer science. One good practice is to have a separate marked container for items that do not belong in regular garbage. Beyond that, let the custodian know what goes into the garbage can.

Ms. Antal's list will grow, and so will yours. But with practice, safer science will become a habit, just one of the everyday tasks of teaching.

Inventory

Although Mr. Jackson had limited equipment and material, he began thinking of just how much there might be in his entire area. Shelves and cupboards were loaded with "good stuff" kept for when it might be needed. Perhaps it would be a good idea to look around and see just what was stashed away?

Window *Into the* **Classroom**

Establish an area or department inventory. Forms are available, but just a simple list will start the process. Document quantities, and purchase only what you will need for one year. Check disposal requirements; they are often costly and complicated. Be cautious of old or donated items, and don't accept "free" chemicals.

When your inventory is complete, establish a safe storage system. The NSTA safety series, your county health department, and the Flinn catalog are good sources for storage information.

Make sure you have a material safety data sheet (MSDS) for each chemical purchased or kept. Keep and update three sets of the MSDSs: one for your area, one for your principal's office, and one set for your central administration office. Emergency personnel will need them if there is a fire or accident.

Appendix C

Field Trips

Ms Thompson's field trip with her tenth graders went well—in the classroom, on the bus, and at the planetarium. She knew that success was due to careful planning, attention to every detail, being sure that the planetarium staff knew why she would be there, and working productively with the kids.

Window
Into the
Classroom

Field trips, the most challenging and sometimes the most rewarding experiences for teachers and students, are not always a positive experience. Safety issues multiply exponentially the longer a trip lasts. Control of a group of students in a shifting environment is almost impossible. Instead, you must bring your students to the point of ownership: it must be their field trip and their responsibility, too. You may need weeks or even months of preparation.

Think through the planned trip, considering every eventuality. Teach the students about safety issues. Teach them about the safety precautions for where they will be going, for how they will get there, and for what they will be doing. Teach them why. Then consider the academic side of the trip.

Before you announce your plan, talk to your district office about insurance coverage. Does the district policy (Is there one?) cover you and protect you from liability? Then make an appointment with your own insurance agent and attorney. Tell them what you plan to do, and listen carefully to what they have to say.

Permission slips are vital. Most schools now have a uniform slip. If yours does not, work with your district to prepare one.

If any student has a medical condition or allergy, you must know of it and be prepared to cope with a flare-up.

ELL students need extra attention. There may be cultural differences in expected behavior. A student with limited language ability may be retiring and difficult to assess, especially when he or she is confronted with new concepts of participation.

Working Within the System

Mr. Hafiz, known for his success as an ELL teacher, was also known for his success in working within the system. When he found the room ventilation was not functioning

Window
Into the
Classroom

properly, he documented his finding. He met with his principal, explained his concern, and submitted his short written report along with his recommended solution. He knew his request had to go up the line to a district office and then back to the maintenance staff. Until then, he modified his lessons for safer practice.

Mr. Hafiz was persistent when safety was an issue. He had researched federal, state, and local safety regulations, and everyone he worked with knew that. They also knew he would speak plainly, discuss any problem calmly but persistently, and give the district a chance. His proactive approach worked. Everyone understood clearly the personal and district liability if no action were taken on a problem.

The plan is to work together as a team, including all levels of your district, making sure that everyone understands the issues.

Window
Into the
Classroom

Eduardo remembered his science classes with fondness. He had entered school with very limited English proficiency and was now ready to graduate, move on to college, majoring in physics. Careful nurturing and his own participation had allowed him to develop into the success story he now represented.

Eduardo recalled that his teachers had been genuinely interested in him. They had valued him as a person. He knew they had been concerned for his safety, because he remembers the pages his teachers frequently referred to when giving safety instructions, placing check marks here and there.

Eduardo felt fortunate to have found his way to his new country—and to his teachers.

In his mind he thanked them.

Recommended Safety Publications

Kwann, T., and J. Texley. 2002. *Exploring safely: A guide for elementary teachers.* Arlington, VA: NSTA Press.

Kwann, T., and J. Texley. 2003. *Inquiring safely: A guide for middle school teachers.* Arlington, VA: NSTA Press.

Kwann, T., J. Texley , and J. Summers. 2004. *Investigating safely: A guide for high school teachers.* Arlington, VA: NSTA Press.

Internet Resources

A word of caution concerning the internet: accuracy of information is only as good as the sponsoring person or organization. Be sure to check closely and employ a healthy dose of skepticism. If any of these websites have expired, use a search engine to reconnect.

IDEA—The Individuals with Disabilities Education Act. IDEA 1997 at *www.cec.sped.org/law_res/doc/law/index.php.* IDEA 2004 aligns IDEA with No Child Left Behind Act (NCLB). Available at *www.ed.gov/policy/speced/guid/idea/idea2004.html*

ADA—Americans with Disabilities Act available at *www.usdoj.gov/crt/ada/adahom1.htm.* Prohibits discrimination against people with disabilities. Requires access to all facilities.

NIOSH—The National Institute for Occupational Safety and Health at *www.cdc.gov/niosh/homepage.html.*

OSHA—Occupational Safety and Health Administration at *www.osha.gov.*

School Lab Safety Audit, a Washington state checklist for auditing school laboratory safety at *www.k12.wa.us/facilities/healthsafewtyguide.asp.*

Materials Safety Data Sheets (MSDSs), The SIRI MSDS Index with MSDSs for most chemicals available at *http://hazard.com/msds.*

Flinn Chemical and Biological Catalog Reference Manual. Request at *www.flinnsci.com.*

Index

*Page numbers in **boldface** type indicate figures or tables.*

Index

Index

Index

Index

developing process skills of *51–55*, **53, 54**
directed inquiry *52–53*, **56–57**
full inquiry *55*, **56–57**
guided inquiry *6, 53–55*, **56–57,** *97, 98, 205, 226*
merging SIOP Model and *102–103*
results of *98*
scaffolded *5, 227–228*
shared experiences from *38*
WebQuests for *86–87*
Interdependence of students *48–49*
International Reading Association (IRA) *221*
Internet. *See* World Wide Web
IRA (International Reading Association) *221*

J

Jordan, M. *17*
Journeys—ELD/ELA in the Content Area: Science *123*

K

K-W-L chart *45*
applications of *113–114, 120–121, 131–132*
Karplus, Robert *16*
Kids URLs *89*
Krashen, Stephen *14, 100*

L

Labeling classroom equipment *40*
Laboratory activities *52. See also* Inquiry-based science activities
adapted lab report for *41*, **42**
cookbook *96, 204*
open-ended *96–98*
Language arts performance chart *29*, **30**
Language functions *51*, **51,** *66–67*, **67**
Language Links *206*
Language minority *227*
Language proficiency *27*, **27,** *172, 227*
standards for (*See* English language proficiency standards)
Lawrence Hall of Science *99*
Learning *5–6*
active cognitive involvement in *16–17, 18*
for all students *5*
assessment of *5, 61–75*
beyond the classroom *6, 79–91*

Index

O

Objectives for learning 5, 6, 68 , 227
Occupational Safety and Health Administration (OSHA) 230
On Our Way to English 111, 112
Online Directory of ESL Resources 222
OSHA (Occupational Safety and Health Administration) 230

P

P-SELL (Promoting Science Among English Language Learners) (Florida) 214
Parental involvement 83
   in Family Science program 85–86
   in science fairs 84–85
Parental permission for field trips 232
Pathways to the Science Education Standards 182
Patterns of scientific discourse 50, **50,** 207–208
Performance indicators 190, 195, 227
Performance standards 213, 227
Permission slips for field trips 232
Pestalozzi, Henrich 15
Piaget, Jean 15, 16
Planetarium 80, 81
Planning
   for assessment 68–73
     collecting and recording student information 69–71, **71**
     components for **70**
     identifying learning standards 68–69
     scoring criteria and data interpretation 71–72, **72**
     steps in **68**
     use of data and reporting results 72–73
   for instruction 25–35
     putting pieces together for 31–33, **32, 34**
     using standards for 26–28, **27,** 31–32
     using student information for 28–31, **30**
Plants: kindergarten (lesson plan) 111–114
Primary language fluency 28, 207
Primary language support 12, 30, 41, 227
Problem-centered learning approach 16
Process skills of inquiry 51–55, **53, 54,** 228
Professional development 7
Professional organizations and resources 222–223
Project 2061 180, 181
Promoting Science Among English Language Learners (P-SELL) (Florida) 214
Pull-out model 12, 227

Index